SpringerBriefs in Statistics

JSS Research Series in Statistics

Editors-in-Chief

Naoto Kunitomo, The Institute of Statistical Mathematics, Tachikawa-shi, Japan

Akimichi Takemura, The Center for Data Science Education and Research, Shiga University, Hikone, Japan

Series Editors

Shigeyuki Matsui, Graduate School of Medicine, Nagoya University, Nagoya, Japan

Manabu Iwasaki, School of Data Science, Yokohama City University, Yokohama, Japan

Yasuhiro Omori, Graduate School of Economics, The University of Tokyo, Bunkyo-ku, Japan

Masafumi Akahira, Institute of Mathematics, University of Tsukuba, Tsukuba, Japan

Masanobu Taniguchi, School of Fundamental Science and Engineering, Waseda University, Shinjuku-ku, Japan

Hiroe Tsubaki, The Institute of Statistical Mathematics, Tachikawa, Japan

Satoshi Hattori, Faculty of Medicine, Osaka University, Suita, Japan

Kosuke Oya, School of Economics, Osaka University, Toyonaka, Japan

Taiji Suzuki, School of Engineering, University of Tokyo, Tokyo, Japan

Kunio Shimizu, The Institute of Statistical Mathematics, Tachikawa, Japan

The current research of statistics in Japan has expanded in several directions in line with recent trends in academic activities in the area of statistics and statistical sciences over the globe. The core of these research activities in statistics in Japan has been the Japan Statistical Society (JSS). This society, the oldest and largest academic organization for statistics in Japan, was founded in 1931 by a handful of pioneer statisticians and economists and now has a history of about 90 years. Many distinguished scholars have been members, including the influential statistician Hirotugu Akaike, who was a past president of JSS, and the notable mathematician Kiyosi Itô, who was an earlier member of the Institute of Statistical Mathematics (ISM), which has been a closely related organization since the establishment of ISM. The society has two academic journals: the *Japanese Journal of Statistics and Data Science* (JJSD, Springer), which is the successor of the *Journal of the Japan Statistical Society* (JJSS) and the *Journal of the Japan Statistical Society* (Japanese Series). The membership of JSS consists of researchers, teachers, and professional statisticians in many different fields including mathematics, statistics, engineering, medical sciences, government statistics, economics, business, psychology, education, and many other natural, biological, and social sciences. The JSS Series of Statistics aims to publish recent results of current research activities in the areas of statistics and statistical sciences in Japan that otherwise would not be available in English; they are complementary to the two JSS academic journals, both English and Japanese. Because the scope of a research paper in academic journals inevitably has become narrowly focused and condensed in recent years, this series is intended to fill the gap between academic research activities and the form of a single academic paper. The series will be of great interest to a wide audience of researchers, teachers, professional statisticians, and graduate students in many countries who are interested in statistics and statistical sciences, in statistical theory, and in various areas of statistical applications.

Akihiko Takahashi · Toshihiro Yamada

Asymptotic Expansion and Weak Approximation

Applications of Malliavin Calculus and Deep Learning

Akihiko Takahashi
Graduate School of Economics
The University of Tokyo
Bunkyo-ku, Tokyo, Japan

Toshihiro Yamada
Graduate School of Economics
Hitotsubashi University
Kunitachi-shi, Tokyo, Japan

ISSN 2191-544X   ISSN 2191-5458 (electronic)
SpringerBriefs in Statistics
ISSN 2364-0057   ISSN 2364-0065 (electronic)
JSS Research Series in Statistics
ISBN 978-981-96-8279-9   ISBN 978-981-96-8280-5 (eBook)
https://doi.org/10.1007/978-981-96-8280-5

© The Editor(s) (if applicable) and The Author(s), under exclusive license to Springer Nature Singapore Pte Ltd. 2025

This work is subject to copyright. All rights are solely and exclusively licensed by the Publisher, whether the whole or part of the material is concerned, specifically the rights of translation, reprinting, reuse of illustrations, recitation, broadcasting, reproduction on microfilms or in any other physical way, and transmission or information storage and retrieval, electronic adaptation, computer software, or by similar or dissimilar methodology now known or hereafter developed.
The use of general descriptive names, registered names, trademarks, service marks, etc. in this publication does not imply, even in the absence of a specific statement, that such names are exempt from the relevant protective laws and regulations and therefore free for general use.
The publisher, the authors and the editors are safe to assume that the advice and information in this book are believed to be true and accurate at the date of publication. Neither the publisher nor the authors or the editors give a warranty, expressed or implied, with respect to the material contained herein or for any errors or omissions that may have been made. The publisher remains neutral with regard to jurisdictional claims in published maps and institutional affiliations.

This Springer imprint is published by the registered company Springer Nature Singapore Pte Ltd.
The registered company address is: 152 Beach Road, #21-01/04 Gateway East, Singapore 189721, Singapore

If disposing of this product, please recycle the paper.

To our families

Preface

Asymptotic methods have been widely used in computation or approximation of functions and quantities related to partial differential equations (PDEs), statistics and probability theory in both academics and industry. Particularly, in finance asymptotic expansions for functionals of Brownian motions provide fast and tractable approximations for intractable, but important models in financial markets driven by stochastic differential equations (SDEs). At the beginning of Chap. 7 entitled "Asymptotic Expansion and Weak convergence" in the book of Malliavin and Thalmaier (2006), the authors stated as follows: *In all these developments, the result of Watanabe, which provides the methodology of projecting as asymptotic expansion through a non-degenerated map, plays a key role.* Namely, S. Watanabe introduced the sophisticated theory and computational tool for analyzing Wiener functionals and heat kernels. Since then, based on the result of S. Watanabe asymptotic expansion approaches have been actively developed in computation of expectations on Wiener space within the fields of financial mathematics and statistics for the past three decades, after the earlier studies such as Kunitomo and Takahashi (1992, 2001, 2003), Takahashi (1995, 1999) and Yoshida (1992a,b).

On the other hand, weak approximation of SDEs provides time-discretized computation for expectations or integrals of the solutions of SDEs on Wiener space. Weak approximation has a long history and has been developed by G. Maruyama, G. Milstein and D. Talay and many researchers with a literature of Monte Carlo simulation. Then, at the end of 1990s and the beginning of 2000s, S. Kusuoka introduced a framework of a higher order weak approximation scheme, which works under irregular test functions with a general condition. Today, weak approximation schemes with Monte Carlo methods play important roles in computational mathematics especially in nonlinear problems.

Recently, deep learning methods have been developed as a technique of AI and widely utilized in industries. In applied mathematics, especially in the areas of PDEs and stochastic modeling, neural networks have been used as function approximations (or space-time approximations), which may be regarded as an alternative of finite difference and finite element methods. Since deep learning techniques is generally able to work in high-dimensional settings, it provides a powerful tool in scientific computing.

Our main objective is to "connect" the Watanabe expansion and high order weak approximation of SDEs, which is a continuation of the content of Chap. 7 "Asymptotic Expansion and Weak convergence" in the book of Malliavin and Thalmaier (2006). Concretely, we provide a recent development on asymptotic methods on Wiener space, and then introduce a type of higher order weak approximation of SDEs by certain Brownian polynomials based on asymptotic expansions. Furthermore, another objective of this book is to develop a high order weak approximation scheme with a deep learning method, because it provides wide applications for high-dimensional nonlinear problems. In this regard, we show how to combine our asymptotic expansion based weak approximation with a neural network approximation, which is applicable to high-dimensional nonlinear models.

Chapter 1 and 2 summarize notations and basic facts on probability theory, especially the Itô and Malliavin calculus, respectively. Then, in Chap. 3, Watanabe's asymptotic expansion is reviewed and refined in terms of computational aspects. Chapter 4 provides a general weak approximation scheme based on our expansion method with a numerical recipe. The deep learning application in a high-dimensional nonlinear model is shown in Chap. 5.

The book is written based on a work in JST SAKIGAKE, lecture notes provided in Department of Engineering Science at Osaka University and Department of Mathematics at Kyoto University, and a talk in Bachelier Seminar Paris at H. Poincaré Institute, given by the second author. We are grateful to Professor Masaaki Fukasawa (Osaka University), Professor Emmanuel Gobet (Ecole Polytechnique), Professor Shigeo Kusuoka (University of Tokyo), Professor Takashi Sakajo (Kyoto University) and Professor Jun Sekine (Osaka University) for providing opportunities and motivations for this work. We also thank Professor Riu Naito (University of Toyama) for his continuous support and suggestions on numerical schemes and experiments. Moreover, we greatly appreciate CARF (Center for Advanced Research in Finance) and CIRJE (Center for International Research on the Japanese Economy) in University of Tokyo, CFEE (Center for Financial Engineering Education) in Hitotsubashi University, and GCI Asset Management, Inc. for their constant support of our research. Furthermore, we are grateful to Professor Naoto Kunitomo and Professor Seisho Sato (University of Tokyo) for giving us this opportunity, and we also appreciate Professor Kunitomo for his precious suggestions, which substantially improve the first version of our manuscript. Finally, we would like to thank the Springer staff, particularly, Mr. Praveen Anand Sachidanandam and Mr. Yutaka Hirachi for their support in publishing this book.

We expect that the book will help undergraduate/graduate students, researchers and practitioners who are interested in stochastic calculus, numerical analysis and machine learning to understand the theory and application of asymptotic expansion and weak approximation, as well as to find a new topic in interdisciplinary fields.

Tokyo, Japan
April 2025

Akihiko Takahashi
Toshihiro Yamada

Contents

1 **Introduction** 1
   1.1 Weak Approximation of SDEs 2
   1.2 Deep Learning Approximations 4

2 **Itô Calculus** 7
   2.1 Probability Theory 7
   2.2 Wiener Measure and Brownian Motion 11
   2.3 Itô Integral 13
   2.4 Itô Formula 16
   2.5 SDEs and Diffusion Processes 17

3 **Malliavin Calculus** 21
   3.1 Cameron-Martin Theorem and Elementary IBP Formula 21
   3.2 Malliavin Derivative Operators and Sobolev Spaces 24
   3.3 Skorohod Integral 25
   3.4 Malliavin's IBP Formula 27
   3.5 Watanabe Distributions 30
   3.6 Malliavin Calculus for Multidimensional Diffusions 32

4 **Asymptotic Expansion** 35
   4.1 Asymptotic Expansion of Integrals of Wiener Functionals 36
   4.2 Small Noise Expansion 39
   4.3 Small Time Expansion 41
   4.4 Expansion Around One-Step Euler-Maruyama Scheme 42
   4.5 Explicit Computation and Generalization 45
   4.6 Notes and Summary 52

5 Weak Approximation 55
5.1 From Euler-Maruyama to Higher-Order Weak Approximation 56
5.2 Third-Order Local Approximation 57
5.3 Second-Order Weak Approximation 64
5.4 General High Order Weak Approximation 68
5.5 Numerical Recipe 70
5.6 Notes and Summary 75

6 Application: Deep Learning-Based Weak Approximation 77
6.1 Introduction: Weak Approximation and Partial Differential Equation 77
6.2 Backward Dynamic Programming Principle 79
6.3 Sketch of Weak Approximation Scheme 81
6.4 Deep Neural Network and Universal Approximation 82
6.5 Weak Approximation with Deep Learning 84
6.6 Algorithm, Implementation and Numerical Results 87
6.7 Notes and Summary 90

References 93

Chapter 1
Introduction

In applied mathematics, especially in financial mathematics and AI (Artificial Intelligence), it is significantly important to construct a numerical scheme for a parabolic partial differential equation (PDE)

$$\partial_t u - \mathcal{L} u = 0, \quad u(0, \cdot) = f$$

with a function $f : \mathbb{R}^N \to \mathbb{R}$ and a differential operator $\mathcal{L}$ given by

$$\mathcal{L} = \sum_{j=1}^{N} \sigma_0^j(\cdot) \frac{\partial}{\partial x_j} + \frac{1}{2} \sum_{j_1, j_2=1}^{N} \sum_{i=1}^{d} \sigma_i^{j_1}(\cdot) \sigma_i^{j_2}(\cdot) \frac{\partial^2}{\partial x_{j_1} \partial x_{j_2}},$$

and the corresponding N-dimensional Itô stochastic differential equation (SDE) driven by d-dimensional Brownian motion W:

$$dX_t^x = \sigma_0(X_t^x)dt + \sum_{i=1}^{d} \sigma_i(X_t^x)dW_t^i, \quad X_0^x = x, \tag{1.1}$$

through the relationship $u(t, x) = \mathbb{E}[f(X_t^x)]$, $t > 0$, $x \in \mathbb{R}^N$.

In quantitative finance and generative AI, the solution to the above PDE expressed as the expectation $\mathbb{E}[f(X_t^x)]$ corresponds to the derivative price of an underlying asset X with a payoff f at time t, and the probability distribution function of a target data dynamics X (with a Heaviside function f) at time t, respectively. In such cases, an effective approach to the numerical solution for high-dimensional models is a time-discretization of SDE and computation of the expectation $\mathbb{E}[f(X_t^x)]$ based on Monte Carlo method.

However, it is not easy task to construct a high order discretization scheme if f is not smooth or the diffusion coefficients are not commutative.

Moreover, when considering the computation of the solution to the PDE, the corresponding Monte Carlo method approximates the solution at fixed points t and x rather than computing the entire function u that satisfies the PDE. In particular, for nonlinear PDEs with additional nonlinear terms, solving u numerically requires iterative function approximation of u itself. Moreover, dynamic programming principle formulated such as for American-type option pricing in finance also need nested computation of conditional expectations. While some computational techniques using Monte Carlo methods (such as the least square Monte Carlo method) have been proposed for the problem in the field of stochastic numerical analysis, they involve high computational costs and cannot be applied to high-dimensional models.

Despite these challenges for numerical computation in probability theory and PDEs, recent developments of deep learning enable us to approximate targets on high-dimensional models without suffering from the "curse of dimensionality".

1.1 Weak Approximation of SDEs

In general, when getting the probability distribution function of X_t^x with an unknown solution to the SDE (1.1) at time $t > 0$, we need to construct an approximation scheme $\bar{X}_t^{x,(n)}$ by discretizing the SDE. If the scheme $\bar{X}_t^{x,(n)}$ satisfies

$$|\mathbb{E}[f(X_t^x)] - \mathbb{E}[f(\bar{X}_t^{x,(n)})]| = O\left(\frac{1}{n^m}\right), \tag{1.2}$$

we call it a m-order weak approximation of SDE. When we can simulate independent and identically distributed random variables $\bar{X}_t^{x,(n),[1]}, \ldots, \bar{X}_t^{x,(n),[M]}$, an approximation of the expectation is obtained by Monte Carlo method:

$$\mathbb{E}[f(X_t^x)] \approx \frac{1}{M} \sum_{k=1}^{M} f(\bar{X}_t^{x,(n),[k]}). \tag{1.3}$$

The most fundamental weak approximation method introduced by G. Maruyama is known as the Euler-Maruyama scheme, which is defined as follows: Given an initial value $\bar{X}_0^{\text{EM},x,(n)} = x$,

$$\bar{X}_{kt/n}^{\text{EM},x,(n)} = \bar{X}_{(k-1)t/n}^{\text{EM},x,(n)} + \sigma_0(\bar{X}_{(k-1)t/n}^{\text{EM},x,(n)})\frac{t}{n} + \sum_{i=1}^{d} \sigma_i(\bar{X}_{(k-1)t/n}^{\text{EM},x,(n)})\{W_{kt/n}^i - W_{(k-1)t/n}^i\},$$
$$k = 1, \ldots, n. \tag{1.4}$$

It is easily implemented by simulating Gaussian random variables which describe increments of Brownian motion $W_{kt/n}^i - W_{(k-1)t/n}^i, i = 1, \ldots, d, k = 1, \ldots, n$. The Euler-Maruyama scheme is regarded as the first order weak approximation method satisfying

1.1 Weak Approximation of SDEs

$$|\mathbb{E}[f(X_t^x)] - \mathbb{E}[f(\bar{X}_t^{\text{EM},x,(n)})]| = O\left(\frac{1}{n}\right)$$

for smooth function $f : \mathbb{R}^N \to \mathbb{R}$. Moreover, it works even if f is non-smooth function under a certain ellipticity condition.

How can we construct a weak approximation method of order $m \geq 2$? In principle, a truncated stochastic Taylor expansion scheme obtained by repeatedly applying Itô formula provides a higher order approximation method, where the scheme is shown as

$$\bar{X}_{kt/n}^{x,(n)} = \bar{X}_{(k-1)t/n}^{x,(n)} + \sigma_0(\bar{X}_{(k-1)t/n}^{x,(n)})\frac{t}{n} + \sum_{i=1}^{d} \sigma_i(\bar{X}_{(k-1)t/n}^{x,(n)})\{W_{kt/n}^i - W_{(k-1)t/n}^i\}$$

$$+ \sum_{\ell=2}^{m} \sum_{\alpha_1,\dots,\alpha_\ell=0}^{d} \mathcal{L}_{\alpha_1} \cdots \mathcal{L}_{\alpha_{\ell-1}} \sigma_{\alpha_\ell}(\bar{X}_{(k-1)t/n}^{x,(n)}) \int_{(k-1)t/n < t_1 \cdots < t_\ell < kt/n} dW_{t_1}^{\alpha_1} \cdots dW_{t_\ell}^{\alpha_\ell} \quad (1.5)$$

with notation $dW_t^0 = dt$ and differential operators $\mathcal{L}_0 = \mathcal{L}$ and $\mathcal{L}_i = \sum_{j=1}^{N} \sigma_i^j(\cdot)\frac{\partial}{\partial x_j}$, $i = 1, \dots, d$. However, the problem here is that the probability distribution functions of iterated Itô integrals $\int_{(k-1)t/n < t_1 < \cdots < t_m < kt/n} dW_{t_1}^{\alpha_1} \cdots dW_{t_\ell}^{\alpha_\ell}$, $\alpha \in \{0, 1, \dots, d\}^\ell$, $\ell \leq m$ are unknown, and their random numbers cannot be generated. For example, take $m = 2$ and let $I_{(i,j)}(t) := \int_{0 < t_1 < t_2 < t} dW_{t_1}^i dW_{t_1}^j$ for $i, j = 1, \dots, d$. By Itô formula, one has

$$W_t^i W_t^j = I_{(i,j)}(t) + I_{(j,i)}(t) + t\delta_i^j, \quad (1.6)$$

or equivalently,

$$I_{(i,j)}(t) = \frac{1}{2}W_t^i W_t^j + \frac{1}{2}L_{ij}(t) - \frac{1}{2}t\delta_i^j, \quad (1.7)$$

where δ_j^i is the Kronecker's delta and $L_{ij}(t) = I_{(i,j)}(t) - I_{(j,i)}(t)$ is the Lévy area. Thus, the iterated integrals $I_{(i,j)}(t)$, $i, j = 1, \dots, d$ cannot be obtained by a polynomial of Brownian motion in general, and are obtained if and only if the following commutativity condition holds:

$$\mathcal{L}_i \sigma_j = \mathcal{L}_j \sigma_i, \quad i, j = 1, \dots, d. \quad (1.8)$$

Note that the map $W. \mapsto X_\cdot^x$ (called the Itô map) is continuous if and only if the commutativity condition holds, otherwise it is not continuous (but $\mathbf{W}. \mapsto X_\cdot^x$ is continuous as the Itô-Lyons map where $\mathbf{W}.$ represents a rough path). In summary, higher order schemes cannot be simulated directly with Brownian motion except the case that the Itô map is continuous. This is known as the Lévy area problem. Many methods have been studied for generating random numbers in cases where the commutativity condition is not satisfied. The cost for avoiding the Lévy area problem for all methods in the literature is to use additional random variables whose number is larger than d, i.e. the dimension of Brownian motion, in each time-step.

In the current book, we overcome this problem based on an *asymptotic expansion* approach with Malliavin calculus, of which details will be provided in Chap. 4. More concretely, we first generalize Watanabe's asymptotic expansion under a certain nondegeneracy condition. Namely, with an explicitly calculable coefficient in each order ε^i, we obtain

$$|\mathbb{E}[f(F^\varepsilon)] - \{\mathbb{E}[f(F^0)] + \sum_{i=1}^{m} \varepsilon^i \mathbb{E}[f(F^0)\mathcal{H}_i]\}| = \|f\|_\infty \times O(\varepsilon^{m+1}). \quad (1.9)$$

Then, we construct a m-order weak approximation of SDE with a certain polynomial of Brownian motion by extending Watanabe's asymptotic expansion under ellipticity on diffusion coefficients. Particularly, with the so called Malliavin weight function $\mathcal{M}^{(m)}$, we obtain

$$|\mathbb{E}[f(X_t^x)] - \mathbb{E}[f(\bar{X}_t^{\mathrm{EM},x,(n)}) \prod_{i=1}^{n} \mathcal{M}^{(m)}(t/n, \bar{X}_{(i-1)t/n}^{\mathrm{EM},x,(n)}, W_{it/n} - W_{(i-1)t/n})]| = O\left(\frac{1}{n^m}\right), \quad (1.10)$$

which works even if the commutativity condition does not hold and f is not smooth. Furthermore, the method only requires simulation of Gaussian random variables for increments of Brownian motion whose number is always equal to d in each time-step. The thorough discussion will be given in Chap. 5.

1.2 Deep Learning Approximations

While the scheme in (1.10) can be efficiently computed by Monte Carlo method, it gives only an approximation of $u(t, x) = \mathbb{E}[f(X_t^x)]$ for fixed t and x as in (1.3). Recent developments of deep learning technique provides a spatial approximation of $u(t, \cdot)$ for $t > 0$, that is, we can construct a neural network $u^{\mathcal{N}\mathcal{N}}(t, \cdot)$ approximating the map

$$x \mapsto \mathbb{E}[f(\bar{X}_t^{\mathrm{EM},x,(n)}) \prod_{i=1}^{n} \mathcal{M}^{(m)}(t/n, \bar{X}_{(i-1)t/n}^{\mathrm{EM},x,(n)}, W_{it/n} - W_{(i-1)t/n})]$$

for $t > 0$, which has the form

$$u^{\mathcal{N}\mathcal{N}}(t, \cdot) = (\mathbb{A}_\ell \circ ReLU_{\ell-1} \circ \mathbb{A}_{\ell-1} \circ \cdots \circ ReLU_1 \circ \mathbb{A}_1)(\cdot) \quad (1.11)$$

1.2 Deep Learning Approximations

with appropriate length ℓ, affine transformations $\mathbb{A}_1, \ldots, \mathbb{A}_\ell$ and Rectified Linear Unit functions $ReLU_1, \ldots, ReLU_{\ell-1}$. Then the PDE solution can be approximated as $u(t, \cdot) \approx u^{\mathcal{NN}}(t, \cdot)$.

In Chap. 6, we will show that our weak approximation method together with deep learning is effectively applied to nonlinear problems, especially to pricing complex financial products under high dimensional settings. The proposed deep learning-based weak approximation will be demonstrated with a concrete example, the so-called Bermudan option, that is an option exercisable at pre-specified timings besides maturity date. The details will be discussed in Chap. 6 with the related deep learning studies in the last part of Sect. 6.7.

The organization of this book is as follows: Chaps. 2 and 3 summarize notations and basic facts on probability theory, especially the Itô and Malliavin calculus, respectively. Then, in Chap. 4, Watanabe's asymptotic expansion is reviewed and refined in terms of computational aspects. Chapter 5 provides a general weak approximation scheme based on our expansion method with a numerical recipe. The deep learning application in a high-dimensional nonlinear model is shown in Chap. 6.

Chapter 2
Itô Calculus

2.1 Probability Theory

We give a brief summary of the basics of probability theory, of which details and proofs are found, for instance, in Williams (1991), Malliavin (1995) and Jacod and Protter (2004).

Definition 2.1 (*Measurable space*) Let S be a non-empty set and let $\mathcal{B}$ be a set of subsets such that (i) $S \in \mathcal{B}$, (ii) for $A \in \mathcal{B}$, it holds that $A^c = \{x \in S \; ; \; x \notin A\} \in \mathcal{B}$, (iii) for $\{A_n\}_{n \in \mathbb{N}} \subset \mathcal{B}$, it holds that $\cup_{n \in \mathbb{N}} A_n = \{x \in S \; ; \; \exists i \in \mathbb{N} \text{ s.t. } x \in A_i\} \in \mathcal{B}$. Then $\mathcal{B}$ is called a σ-field and the pair $(S, \mathcal{B})$ is called a measurable space.

Definition 2.2 (*Measure*) Let $(S, \mathcal{B})$ be a measurable space and let $\mu : \mathcal{B} \to [0, \infty]$ be a map satisfying (i) $\mu(\emptyset) = 0$, (ii) for a set of disjoint sets $\{A_n\}_{n \in \mathbb{N}} \subset \mathcal{B}$ (i.e. $A_i \cap A_j = \emptyset, \forall i \neq j$), $\mu(\cup_{n \in \mathbb{N}} A_n) = \sum_{n \in \mathbb{N}} \mu(A_n)$. Then μ is called a measure and the triplet $(S, \mathcal{B}, \mu)$ is called a measure space. If $\mu(S) < \infty$, μ is called a finite measure. If $\exists \{B_n\}_{n \in \mathbb{N}} \subset \mathcal{B}$ s.t. $S = \cup_{n \in \mathbb{N}} B_n$ and $\mu(B_n) < \infty, \forall n \in \mathbb{N}$, μ is called a σ-finite measure.

Definition 2.3 (*Probability measure*) Let $(\Omega, \mathcal{F})$ be a measurable space. Let $P : \mathcal{F} \to [0, 1]$ be a measure such that $P(\Omega) = 1$. We call the map P a probability measure and the triplet $(\Omega, \mathcal{F}, P)$ is called a probability space. If $A \in \mathcal{F}$ satisfies $P(A) = 1$, we say A a.s. (almost surely).

Definition 2.4 (σ-*field generated by a class $\mathcal{A}$ of subsets*) Let S be a set and let $\mathcal{A}(\subset 2^S)$ be a family of subsets (not necessarily σ-field). Then, $\sigma(\mathcal{A})$ given by $\sigma(\mathcal{A}) := \cap_{\mathcal{G}; \; \sigma\text{-field}, \; \mathcal{A} \subset \mathcal{G}} \mathcal{G}$ is the smallest σ-field, in the sense that $\sigma(\mathcal{A})$ is σ-field and if $\mathcal{G}(\supset \mathcal{A})$ is a σ-field, then $\sigma(\mathcal{A}) \subset \mathcal{G}$. We call $\sigma(\mathcal{A})$ the σ-field generated by $\mathcal{A}$.

Definition 2.5 (*Borel field*) Let $(S, \mathcal{O})$ be a topological space. The σ-field $\mathscr{B}(S) := \sigma(\mathcal{O})$ is called the Borel field over S.

Definition 2.6 (*Some σ-fields*) For two σ-fields $\mathcal{G}$ and $\mathcal{H}$, we define a σ-field $\mathcal{G} \vee \mathcal{H} := \sigma(\mathcal{G} \cup \mathcal{H})$. For two σ-fields $\mathcal{G}$ and $\mathcal{H}$, we define a σ-field $\mathcal{G} \otimes \mathcal{H} := \sigma(\{A \times B; \ A \in \mathcal{G}, B \in \mathcal{H}\})$.

Definition 2.7 (*Product measure*) Let $(S_1, \mathcal{B}_1, \mu_1)$ and $(S_2, \mathcal{B}_2, \mu_2)$ be the two σ-finite measure spaces. Then, there exists a unique measure $\lambda : \mathcal{B}_1 \otimes \mathcal{B}_2 \to [0, \infty]$ such that $\lambda(A_1 \times A_2) = \mu_1(A_1) \times \mu_2(A_2), \forall A_i \in \mathcal{B}_i, i = 1, 2$. We write $\mu_1 \otimes \mu_2 = \lambda$ and call it the product measure.

Definition 2.8 (*Independence*) Let $(\Omega, \mathcal{F}, P)$ be a probability space. Let Λ be an index set and let $\mathcal{F}_n, n \in \Lambda$ be sub-σ-fields (i.e. $\mathcal{F}_n \subset \mathcal{F}$ for $n \in \Lambda$). We say that σ-fields $\mathcal{F}_n, n \in \Lambda$ are independent if $P(\cap_{i=1}^N A_i) = \prod_{i=1}^N P(A_i), \forall N \in \mathbb{N}, \forall \lambda_i \in \Lambda, i \leq N, \forall A_i \in \mathcal{F}_{\lambda_i}, i \leq N$.

Definition 2.9 (*Random variable*) Let $(\Omega, \mathcal{F})$ and $(S, \mathcal{B})$ be measurable spaces. A map $X : \Omega \to S$ is called a S-valued random variable if X is a $\mathcal{F}/\mathcal{B}$-measurable function, i.e. for $A \in \mathcal{B}$, $X^{-1}(A) := \{\omega \in \Omega; X(\omega) \in A\} \in \mathcal{F}$.

Definition 2.10 (*σ-field generated by a collection of random variables*) For a random variable $X : \Omega \to \mathbb{R}^N$, we define $\sigma(X) = \sigma(\{X^{-1}(A); A \in \mathscr{B}(\mathbb{R}^N)\})(\equiv \{X^{-1}(A); A \in \mathscr{B}(\mathbb{R}^N)\})$. For a sequence of $\mathbb{R}^N$-valued random variables $\{X_\lambda\}_{\lambda \in \Lambda}$, we write $\sigma(X_\lambda; \lambda \in \Lambda) = \{X_\lambda^{-1}(A); \lambda \in \Lambda, A \in \mathscr{B}(\mathbb{R}^N)\}$.

Theorem 2.1 (Doob-Dynkin) *Let $X : \Omega \to \mathbb{R}^m$ be a random variable. A random variable $Y : \Omega \to \mathbb{R}^n$ is $\sigma(X)$-measurable $\Leftrightarrow \exists \ \mathscr{B}(\mathbb{R}^m)/\mathscr{B}(\mathbb{R}^n)$-measurable map $f : \mathbb{R}^m \to \mathbb{R}^n$ s.t. $Y = f(X)$.*

See Lemma A3.2 (a) in Williams (1991) for the proof of Theorem 2.1.

Definition 2.11 (*Probability distribution*) Let $(\Omega, \mathcal{F}, P)$ be a probability space and $(S, \mathcal{B})$ be a measurable space. For a S-valued random variable $X : \Omega \to S$, the probability distribution (or the probability law) is defined as the measure $P \circ X^{-1} : \mathcal{B} \to [0, 1]$ given by $\mathcal{B} \ni A \mapsto P \circ X^{-1}(A) = P(X^{-1}(A)) = P(\{\omega \in \Omega; X(\omega) \in A\}) \in [0, 1]$.

The relationship of a random variable and its probability distribution is summarized as follows:

$$\begin{array}{ccc} & \Omega \xrightarrow{X} & S \\ [0,1] \xleftarrow{P} & \mathcal{F} \xleftarrow{X^{-1}} & \mathcal{B} \\ [0,1] & \xleftarrow{P \circ X^{-1}} & \mathcal{B} \end{array}$$

Let $(S, \mathcal{B}, \mu)$ be a measure space. Let $\mathcal{S}_+ := \{f : S \to [0, \infty); f = \sum_{i=1}^k a_i \mathbf{1}_{A_i}, k \in \mathbb{N}, a_i \geq 0, i \leq k, \{A_i\}_{i \leq k} \subset \mathcal{B}$ s.t. $A_i \in \mathcal{B}, i \leq k$ are disjoint and $\cup_{i \leq k} A_i = S\}$. For $f \in \mathcal{S}_+$ which has the form $f = \sum_{i=1}^k a_k \mathbf{1}_{A_k}$, define $\int_S f(x) d\mu(x) = \sum_{i=1}^k a_i \mu(A_i)$. For a nonnegative measurable function $f : S \to [0, \infty)$, $\exists \{f_n\}_{n \in \mathbb{N}} \subset \mathcal{S}_+$ s.t. $\lim_{n \to \infty} f_n(x) = f(x), \forall x \in S$, and then the integral is well-defined as $\int_S f(x) d\mu(x) := \lim_{n \to \infty} \int_S f_n(x) d\mu(x) \leq \infty$.

2.1 Probability Theory

Definition 2.12 (*Integral*) Let $f : S \to \mathbb{R}$ be a measurable function, and let $f^+ := \max\{f, 0\}$, $f^- := -\max\{-f, 0\}$ be nonnegative measurable functions. If $\int_S f^+(x)d\mu(x) < \infty$ or $\int_S f^-(x)d\mu(x) < \infty$, $\int_S f(x)d\mu(x) := \int_S f^+(x)d\mu(x) - \int_S f^-(x)d\mu(x) \leq \infty$ is defined. If $\int_S f^+(x)d\mu(x) < \infty$ and $\int_S f^-(x)d\mu(x) < \infty$, then $\int_S f(x)d\mu(x) = \int_S f^+(x)d\mu(x) - \int_S f^-(x)d\mu(x) < \infty$ and f is called an integrable function.

For a random variable X on a probability space $(\Omega, \mathcal{F}, P)$, we write $\mathbb{E}[X] := \int_\Omega X(\omega)dP(\omega)$ if it is well-defined and call it the expectation of X.

Theorem 2.2 (*The elementary formula for expectation*) *Let $(\Omega, \mathcal{F}, P)$ be a probability space. Let $X : \Omega \to \mathbb{R}^N$ be a random variable and $f : \mathbb{R}^N \to \mathbb{R}$ be a measurable function. Then, $f(X(\cdot)) := (f \circ X)(\cdot)$ is integrable on $(\Omega, \mathcal{F}, P)$ if and only if f is integrable on $(\mathbb{R}^N, \mathcal{B}(\mathbb{R}^N), P \circ X^{-1})$, and then it holds that $\mathbb{E}[f(X)] = \int_\Omega f(X(\omega))dP(\omega) = \int_{\mathbb{R}^N} f(x)d(P \circ X^{-1})(x)$.*

See Lemma 6.12 in Williams (1991) for the proof of Theorem 2.2.

Let $(\Omega, \mathcal{F}, P)$ be a probability space. For $p \geq 1$, let $L^p(\Omega, \mathcal{F}, P) = L^p(\Omega) = \{X : \Omega \to \mathbb{R}^N;$ $\mathcal{F}$-measurable map, $\|X\|_p := \mathbb{E}[|X|^p]^{1/p} < \infty\}$ with the usual identification $X = Y$ if and only if $X = Y$ a.s. For $p \geq 1$, $(L^p(\Omega, \mathcal{F}, P), \|\cdot\|_{L^p})$ is a Banach space. If $p = 2$, $(L^2(\Omega), \langle\cdot,\cdot\rangle_{L^2})$ is a Hilbert space where $\langle X_1, X_2\rangle_{L^2} := \mathbb{E}[X_1 X_2]$. More generally, for a measure space $(S, \mathcal{B}, \mu)$, a measurable space $(V, \mathcal{G})$ and a normed space $(V, \|\cdot\|_V)$, for $p \geq 1$, let $L^p(S, \mathcal{B}, \mu) = L^p(S; V) = \{f : S \to V;$ $\mathcal{B}$-measurable map, $\|f\|_{L^p(S;V)} := (\int_S \|f\|_V^p d\mu)^{1/p} < \infty\}$.

Let $(\Omega, \mathcal{F}, P)$ be a probability space.

Definition 2.13 (*Conditional expectation (Kolmogorov)*) Let $\mathcal{G} \subset \mathcal{F}$ be a sub σ-field. For $X \in L^1(\Omega, \mathcal{F}, P)$, $\exists Y \in L^1(\Omega, \mathcal{G}, P)$ s.t. $\mathbb{E}[(X - Y)Z] = 0$, $\forall Z = 1_A$, $A \in \mathcal{G}$, i.e. $\int_A X(\omega)dP(\omega) = \int_A Y(\omega)dP(\omega)$, $\forall A \in \mathcal{G}$. The random variable $Y \in L^1(\Omega, \mathcal{G}, P)$ is called the conditional expectation of X given the σ-field $\mathcal{G}$. We write $Y = \mathbb{E}[X|\mathcal{G}]$, a.s.

If $X \in L^2(\Omega, \mathcal{F}, P)$, then $\mathbb{E}[X|\mathcal{G}] \in L^2(\Omega, \mathcal{G}, P)$ satisfies $\mathbb{E}[(X - \mathbb{E}[X|\mathcal{G}])Z] = \langle X - \mathbb{E}[X|\mathcal{G}], Z\rangle_{L^2} = 0$, $\forall Z \in L^2(\Omega, \mathcal{G}, P)$ and

$$\mathbb{E}[|X - \mathbb{E}[X|\mathcal{G}]|^2] = \inf_{Z \in L^2(\Omega, \mathcal{G}, P)} \mathbb{E}[|X - Z|^2]. \tag{2.1}$$

In the case, the conditional expectation (L^2-projection) $\mathbb{E}[X|\mathcal{G}]$ is called the least squares best estimator. See Chaps. 9.2–9.5 in Williams (1991) and Chap. IV.2 of Malliavin (1995) for the details.

Theorem 2.3 (*Radon-Nikodym theorem*) *Let $(S, \mathcal{B})$ be a measurable space satisfying $\exists \{A_n\}_n \subset \mathcal{B}$ s.t. $S = \cup_n A_n$. Let μ and ν be two σ-finite measures on $(S, \mathcal{B})$. We say*

ν is absolutely continuous with respect to μ if for $A \in \mathcal{F}$, $\mu(A) = 0 \Rightarrow \nu(A) = 0$. If ν is absolutely continuous with respect to μ, then $\exists$ integrable $\frac{d\nu}{d\mu} : S \to \mathbb{R}^+$ s.t. $\nu(A) = \int_A \frac{d\nu}{d\mu}(x) d\mu(x)$, $\forall A \in \mathcal{B}$. The function $\frac{d\nu}{d\mu}$ is called the Radon-Nikodym derivative of ν with respect to μ. We say μ and ν are equivalent if μ and ν are absolutely continuous each other, i.e. for $A \in \mathcal{B}$, $\mu(A) = 0 \Leftrightarrow \nu(A) = 0$.

See Chaps. 14.13–14.16 in Williams (1991) and Chap. IV.6 of Malliavin (1995) for the details.

Definition 2.14 (*Probability density function*) Let $(\Omega, \mathcal{F}, P)$ be a probability space. Let $X : \Omega \to \mathbb{R}^N$ be a random variable and suppose that $P \circ X^{-1}$ is absolutely continuous with the Lebesgue measure Leb on $(\mathbb{R}^N, \mathcal{B}(\mathbb{R}^N))$. Then, $p^X(x) := \frac{dP \circ X^{-1}}{d\text{Leb}}(x) =: \frac{dP \circ X^{-1}(x)}{dx}$, $x \in \mathbb{R}^N$ is called the probability density function of X.

Definition 2.15 (*Convergences*) Let $(\Omega, \mathcal{F}, P)$ be a probability space.

1. We say $\{X_n\}_{n \in \mathbb{N}}$ converges to X in probability if $\forall \epsilon > 0$, $\lim_{n \to \infty} P(\{\omega \in \Omega; |X_n(\omega) - X(\omega)| \geq \epsilon\}) = 0$ (i.e. $\forall \epsilon > 0$, $\forall \epsilon' > 0$, $\exists N = N(\epsilon') \in \mathbb{N}$ s.t. $\forall n \geq N$, $P(|X_n - X| \geq \epsilon) < \epsilon'$). We write $X_n \to X$ in prob.
2. For $p \geq 1$, we say $\{X_n\}_{n \in \mathbb{N}}$ converges to X in L^p if $\lim_{n \to \infty} \mathbb{E}[|X_n - X|^p] = 0$. We write $X_n \to X$ in $L^p(\Omega)$.
3. We say $\{X_n\}_{n \in \mathbb{N}}$ converges to X almost surely (a.s.) if $P(\{\omega \in \Omega; \lim_{n \to \infty} X_n(\omega) = X(\omega)\}) = P(\cap_{p \geq 1} \cup_{N \geq 1} \cap_{m \geq N} \{\omega; |X_m(\omega) - X(\omega)| < 1/p\}) = 1$.
4. We say $\{X_n\}_{n \in \mathbb{N}}$ converges to X in distribution (or in law) if $\lim_{n \to \infty} \mathbb{E}[f(X_n)] = \mathbb{E}[f(X)]$ for any bounded continuous function f.

The convergence in distribution is also called weak convergence. Note that in weak convergence, it is not necessary to assume that $\{X_n\}_{n \in \mathbb{N}}$ and X are defined on the same probability space in general, while in almost sure or L^p-convergence, or convergence in probability, one needs that $\{X_n\}_{n \in \mathbb{N}}$ and X are all defined on the same probability space. See Chaps. 17 and 18 of Jacod and Protter (2004).

Let $(\Omega, \mathcal{F}, P)$ be a probability space and let $\mathbf{T}$ be a time index set $[0, \infty)$ (or $\mathbb{N} \cup \{0\}$).

Definition 2.16 (*Stochastic process*) A family $\{X_t\}_{t \in \mathbf{T}}$ is called a stochastic process if $\forall t \in \mathbf{T}$, $X_t : \Omega \ni \omega \mapsto X_t(\omega) \in \mathbb{R}^N$ is a random variable. For a fixed $\omega \in \Omega$, $\mathbf{T} \ni t \mapsto X_t(\omega)$ is called a *path*.

Definition 2.17 (*Filtration*) A family of sub σ-fields $\{\mathcal{F}_t\}_{t \in \mathbf{T}}$ is called a filtration if $\mathcal{F}_s \subset \mathcal{F}_u$ holds for all $s \leq u$.

Let $\{\mathcal{F}_t\}_{t \in \mathbf{T}}$ be a filtration.

Definition 2.18 (*Adapted process*) A stochastic process $\{X_t\}_{t \in \mathbf{T}}$ is adapted to the filtration $\{\mathcal{F}_t\}_{t \in \mathbf{T}}$ or a $\{\mathcal{F}_t\}$-adapted process if X_t is $\mathcal{F}_t$-measurable for each $t \in \mathbf{T}$, i.e. $\{\omega; X_t(\omega) \in A\} \in \mathcal{F}_t$, $\forall A \in \mathcal{B}(\mathbb{R}^N)$.

2.2 Wiener Measure and Brownian Motion

Definition 2.19 (*Markov process*) A stochastic process $\{X_t\}_{t\in \mathbf{T}}$ is called a Markov process if (i) $\{X_t\}_t$ is $\{\mathcal{F}_t\}$-adapted, (ii) $\mathbb{E}[f(X_{t+s})|\mathcal{F}_s] = \mathbb{E}[f(X_{t+s})|\sigma(X_s)]$ a.s. $\forall s, t \in \mathbf{T}, \forall$ bounded measurable $f : \mathbb{R}^N \to \mathbb{R}$.

Definition 2.20 (*Martingele*) A stochastic process $\{X_t\}_{t\in \mathbf{T}}$ is called a $\{\mathcal{F}_t\}$-martingale if (i) $\{X_t\}_t$ is $\{\mathcal{F}_t\}$-adapted, (ii) $\mathbb{E}[|X_t|] < \infty, \forall t \in \mathbf{T}$, (iii) $\mathbb{E}[X_{t+s}|\mathcal{F}_s] = X_s$ a.s. $\forall s, t \in \mathbf{T}$.

See Appendix A of Nualart and Nualart (2018) for basics of stochastic processes.

In the book, we use the following notations.

- $\lfloor x \rfloor$; the integer part of $x \in \mathbb{R}$.
- $\mathscr{B}_b(\mathbb{R}^N) := \{f : \mathbb{R}^N \to \mathbb{R}; \ f \text{ is bounded measurable function}\}$.
- $\|f\|_\infty := \sup_{x\in\mathbb{R}^N} |f(x)|$ for $f \in \mathscr{B}_b(\mathbb{R}^N)$.
- $C(E) := \{f : E \to \mathbb{R}; \ f \text{ is continuous function where } E \text{ is a topological space}\}$.
- $C_b^k(\mathbb{R}^N) := \{f : \mathbb{R}^N \to \mathbb{R}; \ f \text{ is } k\text{-times continuous differentiable function s.t. } f$ and its derivatives of orders up to k are bounded$\}$.
- $C_b^\infty(\mathbb{R}^N; \mathbb{R}^d) := \{f : \mathbb{R}^N \to \mathbb{R}^d; \ f \text{ is infinitely differentiable function s.t. } f$ and its derivatives of all orders are bounded$\}$.
- $C_b^\infty(\mathbb{R}^N) := C_b^\infty(\mathbb{R}^N; \mathbb{R})$.
- $C^{1,2}([0, T] \times \mathbb{R}) := \{f : [0, T] \times \mathbb{R} \ni (t, x) \mapsto f(t, x) \in \mathbb{R};$ f is twice differentiable in x and once in $t\}$.
- $\mathcal{N}(\mu, \Sigma)$; Gaussian distribution with mean vector μ and covariance matrix Σ.

In the remaining sections of this chapter, we provide basic results for stochastic calculus, of which details and proofs are found, for instance, in Ikeda and Watanabe (1989), Karatzas and Shreve (1991), Rogers and Williams (1994), Funaki (1997), Kusuoka (2020).

2.2 Wiener Measure and Brownian Motion

Definition 2.21 (*Brownian motion*) A stochastic process $\{W_t\}_{0\leq t\leq T}$ defined on a probability space $(\Omega, \mathcal{F}, P)$ is a d-dimensional Brownian motion if

1. $W_0 = 0$, $t \mapsto W_t$ is continuous a.s.
2. for $0 \leq s < t$, $W_t - W_s \sim \mathcal{N}(0, (t-s)I_d)$,
3. for $0 = t_0 < t_1 < \ldots < t_n$, $W_{t_1}, W_{t_2} - W_{t_1}, \ldots, W_{t_n} - W_{t_{n-1}}$ are independent.

The existence of Brownian motion on such a probability space was proved by N. Wiener in 1923 (Wiener 1923). Let $\mathcal{W} := C_{(0)}([0, T]; \mathbb{R}^d) := \{\omega : [0, T] \to \mathbb{R}^d; \ \omega(0) = 0, \omega \text{ is continuous }\}$. For $\omega \in \mathcal{W}$, let $\|\omega\|_\mathcal{W} := \sup_{t\in[0,T]} |\omega(t)|$. Let $\mathscr{O}$ be the set of open sets induced by the distance $d_\mathcal{W}(x, y) = \|x - y\|_\mathcal{W}$ and let $\mathscr{B}(\mathcal{W}) = \sigma(\mathscr{O})$. Here, $\mathscr{B}(\mathcal{W})$ corresponds to the σ-field $\sigma(\{\{\omega \in \mathcal{W}; \ \omega(t_i) \in$

A_i, $i = 1, \ldots, n\}$; $0 < t_1 < \cdots < t_n$, $A_1, \ldots, A_n \in \mathcal{B}(\mathbb{R}^d)$, $n \geq 1\})$. We note that $(\mathcal{W}, \|\cdot\|_{\mathcal{W}})$ is an infinite dimensional Banach space.

The celebrated theorem by N. Wiener is as follows.

Theorem 2.4 (Wiener) *There is a probability space on which Brownian motion can be defined, i.e. on the measurable space $(\mathcal{W}, \mathcal{B}(\mathcal{W}))$, there exists a measure $\mu : \mathcal{B}(\mathcal{W}) \to [0, 1]$ s.t.*

1. $\mu(\mathcal{W}) = 1$,
2. *for $0 \leq s < t$ and $A \in \mathcal{B}(\mathbb{R}^d)$,*

$$\mu(\{\omega \in \mathcal{W};\ \omega(t) - \omega(s) \in A\}) = \int_A \left(\frac{1}{2\pi(t-s)}\right)^{d/2} \exp\left(-\frac{|x|^2}{t-s}\right) dx,$$

3. *for $0 < t_1 < \cdots < t_n$ and $A_i \in \mathcal{B}(\mathbb{R}^d)$, $i = 1, \ldots, n$,*

$$\mu(\{\omega \in \mathcal{W};\ \omega(t_i) - \omega(t_{i-1}) \in A_i,\ 1 \leq i \leq n\})$$
$$= \prod_{i=1}^{n} \mu(\{\omega \in \mathcal{W};\ \omega(t_i) - \omega(t_{i-1}) \in A_i\}).$$

Proof See Theoreom I.7.1 in Ikeda and Watanabe (1989), and Theorem I.6.1 in Rogers and Williams (1994). □

The measure μ is called the Wiener measure. A d-dimensional Brownian motion is given through the coordinate map $W. : \Omega \ (:= \mathcal{W}) \ni \omega \mapsto W.(\omega) = \omega \in \mathcal{W}$, i.e.

$$W_t(\omega) = \omega(t) \text{ for } \omega \in \Omega \text{ and } t \in [0, T],$$

whose law is μ, on a probability space $(\Omega, \mathcal{F}, P)$ where $\Omega = \mathcal{W}$, $\mathcal{F} = \mathcal{B}(\mathcal{W})$ and $P = \mu$. We call $\{\mathcal{F}_t\}_{t \geq 0}$ the Brownian filtration if $\mathcal{F}_t := \sigma(W_s; s \leq t) \vee \mathcal{N}$, $t \in [0, T]$, where $\mathcal{N} = \{N;\ P(N) = 0\}$. Then, we easily obtain the following.

Theorem 2.5 $\{W_t\}_{t \geq 0}$ *is a $\{\mathcal{F}_t\}$-martingale.*

Moreover, we have the following results.

Theorem 2.6 $\{W_t\}_{t \geq 0}$ *is a Markov process.*

Proof See Theorem 2.6.15 in Karatzas and Shreve (1991). □

Theorem 2.7 $t \mapsto W_t$ *is not differentiable a.s.*

Proof See Theorem 2.9.18 in Karatzas and Shreve (1991). □

For a function $g : [0, t] \to \mathbb{R}^d$ and a partition $\pi \in \mathcal{P}_{[0,t]} = \{\{t_0, t_1, \ldots, t_n\}; 0 = t_0 < t_1 < \cdots < t_n = t\}$, let $V^{(2)}(g, \pi) := \sum_{i=1}^{n} |g(t_i) - g(t_{i-1})|^2$ be the quadratic variation. For $\pi \in \mathcal{P}_{[0,t]}$, we define $|\pi| = \sup_i |t_i - t_{i-1}|$.

Theorem 2.8 *(i) The quadratic variation of W on $[0, t]$ is t, i.e.*

$$\lim_{|\pi| \to 0} V^{(2)}(W, \pi_{[0,t]}) = t \quad \text{in } L^2(\Omega).$$

(ii) The path $t \mapsto W_t$ is not bounded variation function a.s., i.e.

$$\sup_{\pi \in \mathcal{P}_{[0,T]}} \sum_{i=1}^n |W_{t_i} - W_{t_{i-1}}| = \infty \quad a.s.$$

Proof For (i) see Problem 2.9.8 and its solution in Sect. 2.10 of Karatzas and Shreve (1991). Then, using (i), we easily verify (ii). □

2.3 Itô Integral

If a path $g : [0, T] \to \mathbb{R}^d$ is a differentiable or a bounded variation function, we can define the Riemann-Stieltjes integral "$\int_0^T f(s) dg(s)$" for a continuous function f. However, $W = \{W_t\}_{t \geq 0}$ does not have these properties by the above two theorems, which means that we can not define the integral with respect to Brownian motion $W = \{W_t\}_{t \geq 0}$ in the Riemann-Stieltjes sense.

K. Itô defines a type of integral $\int_0^T f(s) dW_s$ for a stochastic process $\{f(s)\}_{0 \leq s \leq T}$ called *Itô integral* using L^2-theory so that we can rigorously treat a stochastic differential equation $dX_t = \sigma(X_t) dW_t$ driven by Brownian motion.

Let $\{W_t\}_{t \geq 0}$ be a d-dimensional Brownian motion on a probability space $(\Omega, \mathcal{F}, P)$ and let $\{\mathcal{F}_t\}_t$ be the Brownian filtration. We define a class of stochastic processes.

Definition 2.22 (L_a^2: *a class of square-integrable adapted processes*) Let

$$L^2([0, T] \times \Omega; \mathbb{R}^d) := \{X = \{X_t\}_t : [0, T] \times \Omega \to \mathbb{R}^d;$$

$$X \text{ is } \mathscr{B}([0, T]) \otimes \mathcal{F}\text{-measurable}, \int_{[0,T] \times \Omega} |X_t(\omega)|^2 d(\text{Leb} \otimes P)(t, \omega) < \infty\},$$

$$L_a^2 := \{X = \{X_t\}_t \in L^2([0, T] \times \Omega; \mathbb{R}^d); X \text{ is } \{\mathcal{F}_t\}\text{-adapted}\}.$$

We remark that $f \in L_a^2$ can be regarded as a (version of) progressively measurable process (see p. 55 of Funaki (1997)).

For $X, Y \in L_a^2$, we define $\langle X, Y \rangle_{L_a^2} := \mathbb{E}[\int_0^T X_s Y_s ds]$. Then, $(L_a^2, \langle \cdot, \cdot \rangle_{L_a^2})$ is a Hilbert space, i.e. a complete metric space under the distance:

$$d_{L_a^2}(X, Y) := \langle X - Y, X - Y \rangle_{L_a^2}^{1/2} = \|X - Y\|_{L^2([0,T] \times \Omega)} = \mathbb{E}[\int_0^T |X_s - Y_s|^2 ds]^{1/2}.$$

We define Itô integral $I_T(X) = \int_0^T X_s dW_s$ for $X \in L_a^2$ as a map

$$I_T : L_a^2 \ni X \mapsto I_T(X) \in L^2(\Omega)$$

as follows:

1. For an element X in a class of simple predictable processes $\mathcal{P}(\overset{\text{dense}}{\subset} L_a^2)$, we define elementary Itô integral $I_T(X)$ as the sum of "$\mathcal{F}_{t_{i-1}}$-measurable function" × "increment of Brownian motion $(W_{t_i} - W_{t_{i-1}})$" which satisfies the isometry property $\|X\|_{L^2([0,T]\times\Omega)} = \|I(X)\|_{L^2(\Omega)}$ for $X \in \mathcal{P}$.
2. For a general square-integrable adapted process $X \in L_a^2$, taking a sequence such that $\lim_n \|X^n - X\|_{L^2([0,T]\times\Omega)} = 0$, we define the Itô integral $I_T(X) \in L^2(\Omega)$ by using the facts that both L_a^2 and $L^2(\Omega)$ are Hilbert spaces:

$$\begin{array}{ccc} \mathcal{P} \ni X^n & \overset{\text{isometry}}{\longmapsto} & I_T(X^n) \in L^2(\Omega) \\ \cap & \downarrow & \downarrow \\ L_a^2 \ni X & & \exists I_T(X) \in L^2(\Omega). \end{array}$$

Definition 2.23 ($\mathcal{P}$: *a class of simple predictable processes*) Let $\mathcal{P}$ be a subset of L_a^2 defined by

$$\mathcal{P} := \Big\{ \{X_s\}_s \in L_a^2; \ (s,\omega) \mapsto X_s(\omega)$$
$$= \sum_{i=1}^n \xi_{i-1}(\omega) \mathbf{1}_{[t_{i-1},t_i)}(s), \ n \in \mathbb{N}, \ 0 = t_0 < \cdots < t_n = T,$$
$$\xi_{i-1} \text{ is } \mathbb{R}^d \text{-valued } \mathcal{F}_{t_{i-1}}$$
-measurable random variable and $\mathbb{E}[|\xi_{i-1}|^2] < \infty, \ 1 \le i \le n \Big\}.$

Definition 2.24 (*Elementary Itô integral for* $\mathcal{P}$)
For $X \in \mathcal{P}$ with the form $(s,\omega) \mapsto X_s(\omega) = \sum_{i=1}^n \xi_{i-1}(\omega) \mathbf{1}_{[t_{i-1},t_i)}(s)$, we define $I_T(X)$ by

$$\omega \mapsto I_T(X)(\omega) := \sum_{i=1}^n \xi_{i-1}(\omega)[W_{t_i}(\omega) - W_{t_{i-1}}(\omega)].$$

Then, we easily have the following.

Lemma 2.1 (*Properties of Itô integral for* $X \in \mathcal{P}$)

1. (*Isometry*) $\|I_T(X)\|_{L^2(\Omega)} = \|X\|_{L^2([0,T]\times\Omega)}, \ \forall X \in \mathcal{P}$.
2. (*Linearity*) $I_T(c_1 X_1 + c_2 X_2) = c_1 I_T(X_1) + c_2 I_T(X_2), \ \forall c_1, c_2 \in \mathbb{R}, \ \forall X_1, X_2 \in \mathcal{P}$.

We have "$\mathcal{P} \overset{\text{dense}}{\subset} L_a^2$".

2.3 Itô Integral

Lemma 2.2 $\mathcal{P}$ is a dense subset of L_a^2, i.e. $\forall X \in L_a^2$, $\exists \{X^n\}_{n \in \mathbb{N}} \subset \mathcal{P}$ s.t. $\lim_{n \to \infty} \|X - X^n\|_{L^2([0,T] \times \Omega)} = 0$.

Proof See Lemma II.1.1 in Ikeda and Watanabe (1989). □

Let $X \in L_a^2$. We can take a sequence of simple predictable processes $\{X^n\}_{n \in \mathbb{N}} \subset \mathcal{P}$ such that $\|X^n - X\|_{L^2([0,T] \times \Omega)} \to 0$ by the above lemma. Then, $\{X^n\}_{n \in \mathbb{N}} \subset \mathcal{P}$ becomes a Cauchy sequence in L_a^2. Using the isometry and the linearity of elementary Itô integral for $\mathcal{P}$, we have

$$\|I_T(X^m) - I_T(X^n)\|_{L^2(\Omega)}$$
$$= \|I_T(X^m - X^n)\|_{L^2(\Omega)} = \|X^m - X^n\|_{L^2([0,T] \times \Omega)} \to 0, \; n, m \to \infty.$$

Therefore, $\{I_T(X^n)\}_{n \in \mathbb{N}}$ is a Cauchy sequence in $L^2(\Omega)$. Since $L^2(\Omega)$ is a Hilbert space, we have

$$\exists I_T(X) \in L^2(\Omega) \; s.t. \; \lim_{n \to \infty} \|I_T(X) - I_T(X^n)\|_{L^2(\Omega)} = 0. \tag{2.2}$$

Definition 2.25 (Itô integral) For $X \in L_a^2$, the random variable $I_T(X) \in L^2(\Omega)$ defined by (2.2) is called the Itô integral. Such variable $I_T(X)$ is well-defined, i.e. the limit $I_T(X)$ is independent of the choice of a sequence $\{X^n\}_n \subset \mathcal{P}$ such that $X^n \to X$ in L_a^2. We write $\int_0^T X_s dW_s := I_T(X)$.

Theorem 2.9 (Properties of Itô integral) Let $X, Y \in L_a^2$. Then we have $\mathbb{E}[\int_0^T X_s dW_s] = 0$. Also, it holds that $\mathbb{E}[\int_0^T X_s dW_s \int_0^T Y_s dW_s] = \mathbb{E}[\int_0^T X_s Y_s ds]$, in other words, the following isometry (Itô's isometry) holds for the map $I_T : L_a^2 \ni X \mapsto I_T(X) \in L^2(\Omega)$:

$$\mathbb{E}[\int_0^T X_s^2 ds]^{1/2} = \mathbb{E}[(\int_0^T X_s dW_s)^2]^{1/2}$$
$$\| \qquad\qquad\qquad \|$$
$$\|X\|_{L^2([0,T] \times \Omega)} = \|I_T(X)\|_{L^2(\Omega)}.$$

Proof See Proposition II.1.1 in Ikeda and Watanabe (1989). □

Remark 2.1 Similarly, we can define Itô integral "process" as a continuous L^2-martingale (through martingale transform) as follows:

$$\begin{array}{ccc} \mathcal{P} \ni X^n & \stackrel{\text{isometry}}{\longmapsto} & I(X^n) = \{I_t(X^n)\}_t \in \mathcal{M}_T^2 \\ \cap & \downarrow & \downarrow \\ L_a^2 \ni X & \stackrel{\text{isometry}}{\longmapsto} & I(X) = \{I_t(X)\}_t \in \mathcal{M}_T^2. \end{array}$$

In the figure, a Hilbert space $\mathcal{M}_T^2$ denotes $\mathcal{M}_T^2 := \{M = \{M_t\}_{t \leq T}; \; M \text{ is square integrable continuous martingale }\}$. Furthermore, we can extend the domain of the stochastic processes and define Itô integral process $I(X) = \{I_t(X)\}_t$ for $X \in L^2_{loc}(\supset L^2_a)$ where $L^2_{loc} = \{X : [0, T] \times \Omega \to \mathbb{R}^d; \mathscr{B}([0, T]) \otimes \mathcal{F}$-measurable, $\{\mathcal{F}_t\}$-adapted, $\int_0^T |X(t)|^2 dt < \infty$ a.s. $\}$. In such case, $I(X)$ is not a martingale but a local martingale. We can construct Itô integral process as follows:

$$L^2_{loc} \ni X \mapsto I(X) = \{I_t(X)\}_t \in \mathscr{M}_T^{2,loc}$$

where $\mathscr{M}_T^{2,loc} := \{M = \{M_t\}_{t \leq T}; M$ is square integrable continuous local martingale $\}$. In the case, the stochastic integral is determined in the sense of convergence in probability.

Remark 2.2 If $X \in L^2_{loc}$ is a continuous process, we have

$$\int_0^T X_s dW_s = \lim_{|\pi| \to 0} \sum_{k=1}^n X_{t_{k-1}}(W_{t_k} - W_{t_{k-1}}) \text{ in prob}$$

where $\pi \in \{\{t_0, t_1, \ldots, t_n\}; \quad 0 = t_0 < t_1 < \cdots < t_n = T\}$ and $|\pi| := \sup_{k=1,\ldots,n}(t_k - t_{k-1})$.

2.4 Itô Formula

Let $W = \{(W_t^1, \ldots, W_t^d)\}_t$ be a d-dimensional Brownian motion and let $\{\mathcal{F}_t\}_t$ be the Brownian filtration.

For the case $d = 1$, K. Itô initially introduced the following formula in 1942 (cf. Sect. 7 in Itô 1942).

Theorem 2.10 (Itô formula (K. Itô)) *Let* $f \in C_b^2(\mathbb{R})$. *Then,*

$$f(W_t) = f(W_0) + \int_0^t f'(W_s) dW_s + \frac{1}{2} \int_0^t f''(W_s) ds, \ t \geq 0 \ a.s.$$

The formula can be extended to general Itô processes.

Theorem 2.11 (Itô formula for Itô process) *Assume that* $X_0^i \in \mathbb{R}$, b^i *is an adapted process such that* $\int_0^T |b_s^i| ds < \infty$ *a.s. and* $\sigma_{j,\cdot}^i \in L^2_{loc}$ *for* $i = 1, \ldots, N$ *and* $j = 1, \ldots, d$. *Define a stochastic process* $\{X_t\}_{t \geq 0}$ *called Itô process given by*

$$X_t^i = X_0^i + \int_0^t b_s^i ds + \sum_{j=1}^d \int_0^t \sigma_{j,s}^i dW_s^j, \ 0 \leq t \leq T, i = 1, \ldots, N \ a.s.$$

which is formally written in differential form $dX_t = b_t dt + \sum_{j=1}^d \sigma_{j,t} dW_t^j$. *For* $f \in C_b^{1,2}([0, T] \times \mathbb{R}^N)$, *we have*

$$f(t, X_t) = f(0, X_0) + \int_0^t \frac{\partial}{\partial t} f(s, X_s) ds + \sum_{i=1}^N \int_0^t \frac{\partial}{\partial x_i} f(s, X_s) b_s^i ds$$
$$+ \sum_{i=1}^N \sum_{j=1}^d \int_0^t \frac{\partial}{\partial x_i} f(s, X_s) \sigma_{j,s}^i dW_s^j + \frac{1}{2} \sum_{i_1,i_2=1}^N \sum_{j=1}^d \int_0^t \frac{\partial^2}{\partial x_{i_1} \partial x_{i_2}} f(s, X_s) \sigma_{j,s}^{i_1} \sigma_{j,s}^{i_2} ds,$$
$$t \geq 0 \ a.s.$$

Proof See Theorems 3.3.3 and 3.3.6 in Karatzas and Shreve (1991). $\square$

2.5 SDEs and Diffusion Processes

Let $\{W_t\}_t$ be a d-dimensional Brownian motion on a probability space $(\Omega, \mathcal{F}, P)$ and $\{\mathcal{F}_t\}_t$ be the Brownian filtration.

Definition 2.26 A process $\{X_t\}_t$ is called a solution of a stochastic differential equation (SDE) starting from $x \in \mathbb{R}^N$:

$$dX_t = b(X_t)dt + \sigma(X_t)dW_t \quad (2.3)$$

if $\{X_t\}_t$ has a continuous path, is $\{\mathcal{F}_t\}$-adapted with $\sigma^i(X.) \in L^2_{loc}, b^i(X.) : [0, T] \times \Omega \to \mathbb{R}^N$ s.t. $\int_0^T |b^i(X_s)| ds < \infty$ a.s. and satisfies

$$X_t = x + \int_0^t b(X_s) ds + \int_0^t \sigma(X_s) dW_s, \ t \geq 0 \ a.s.$$

Theorem 2.12 *Let* $b : \mathbb{R}^N \to \mathbb{R}^N$ *and* $\sigma : \mathbb{R}^{N \times d} \to \mathbb{R}^N$ *be Lipschitz continuous functions, i.e.* $\exists C > 0$ *s.t.*

$$|b(x) - b(y)| + \|\sigma(x) - \sigma(y)\| \leq C|x - y|, \ \forall x, y \in \mathbb{R}^N.$$

Then, the unique solution of (2.3) exists.

Proof See IV.3.1 in Ikeda and Watanabe (1989). $\square$

Similarly, we can solve the stochastic differential equation from time $t \geq 0$:

$$X_s = x + \int_t^s b(X_r) dr + \int_t^s \sigma(X_r) dW_r, \ s \geq t,$$

which is adapted to the filtration $\{\mathcal{F}_t\}_t$. Hereafter, we write $X^{t,x}$ for the solution in order to clarify the dependence of the initial point x and time t. If $t = 0$, we may use a notation X^x. For $t \geq s$, the process $\{X_s^{t,\cdot}\}_{s \geq t}$ is called the stochastic flow. It is known that the map $x \mapsto X_s^{t,x}$ is continuous, in particular, if b and σ are smooth, the map $x \mapsto X_s^{t,x}$ becomes smooth (see Kunita 1990). We have the following flow property of the solution.

Theorem 2.13 (Flow property) *For all $t \leq s$ and $x \in \mathbb{R}^N$,*

$$X_s^{0,x} = X_s^{t,X_t^{0,x}} \quad a.s.$$

Proof See Lemma V.13.6 in Rogers and Williams (1994). □

In the following, we check the relationship between SDEs and diffusion processes. Let $X = \{X_t\}_t$ be a solution of the following SDE

$$dX_t = b(X_t)dt + \sum_{i=1}^{d} \sigma_i(X_t)dW_t^i, \quad X_0 = x \in \mathbb{R}^N,$$

where $b : \mathbb{R}^N \to \mathbb{R}^N$ and $\sigma = (\sigma_1, \ldots, \sigma_d) : \mathbb{R}^N \to \mathbb{R}^{N \times d}$ are Lipschitz continuous functions. We define a family of linear operators $\{P_t\}_t$ by $(P_t f)(x) = \mathbb{E}[f(X_t^x)]$ for $t \geq 0, x \in \mathbb{R}^N$ and a bounded measurable function $f : \mathbb{R}^N \to \mathbb{R}$.

Definition 2.27 A continuous Markov process is called a diffusion process.

Theorem 2.14 *The solution of SDE $X = \{X_t\}_t$ is a diffusion process, i.e. X has continuous path with the property: for any bounded measurable function $f : \mathbb{R}^N \to \mathbb{R}$ and $s < t$, $\mathbb{E}[f(X_t)|\mathcal{F}_s] = P_{t-s}f(X_s)$.*

Proof See V.1 and V.3 in Rogers and Williams (1994). □

$\{P_t\}_t$ has the relationship $P_s P_t = P_{s+t}$ which is called the semigroup property obtained through $P_{t+s}f(x) = \mathbb{E}[f(X_{t+s}^x)] = \mathbb{E}[\mathbb{E}[f(X_{t+s}^x)|\mathcal{F}_s]] = \mathbb{E}[P_t f(X_s^x)] = P_s P_t f(x), \forall f \in \mathcal{B}_b(\mathbb{R}^N), \forall x \in \mathbb{R}^N$. The family $\{P_t\}_t$ is called the semigroup.

Definition 2.28 A differential operator $\mathcal{L}$ satisfying $\mathcal{L}\varphi(x) = \frac{d}{dt}\mathbb{E}[\varphi(X_t^x)]|_{t=0}$ for $\varphi \in C_b^2(\mathbb{R}^N)$ is called the generator.

Then, we obtain the following result by using Itô formula.

Proposition 2.1 *The generator $\mathcal{L}$ is given by*

$$\mathcal{L} = \sum_{i=1}^{N} b^i(\cdot)\partial_i + \frac{1}{2}\sum_{i,j=1}^{N}\sum_{k=1}^{d} \sigma_k^i(\cdot)\sigma_k^j(\cdot)\partial_i\partial_j.$$

We give a version of Feynman-Kac formula.

Theorem 2.15 (Feynman-Kac formula) *Let $f \in C(\mathbb{R}^N)$ be a polynomial growth function, and let $u \in C([0,T] \times \mathbb{R}^N) \cap C^{1,2}([0,T) \times \mathbb{R}^N)$ be a polynomial growth function satisfies*

2.5 SDEs and Diffusion Processes

$$(\partial_t + \mathcal{L})u(t,x) = 0, \quad (t,x) \in [0,T) \times \mathbb{R}^N,$$
$$u(T,x) = f(x), \quad x \in \mathbb{R}^N.$$

Then,

$$u(t,x) = P_{T-t}f(x) = \mathbb{E}[f(X^x_{T-t})], \quad (t,x) \in [0,T) \times \mathbb{R}^N.$$

Proof See Theorem 5.7.6 in Karatzas and Shreve (1991). □

Remark 2.3 If the coefficients b, σ and the data f are bounded and smooth with bounded derivatives, then there is a solution u which is bounded and smooth with bounded derivatives. If the coefficients b, σ satisfy the same conditions but f is continuous only, it is not obvious that there always exists a smooth solution u. However, if in addition σ satisfies the uniformly elliptic condition or b, σ satisfy the hypoelliptic condition (weaker than uniformly ellipticity), u may have a smooth property. See Friedman (1964), Kusuoka and Stroock (1984), Kusuoka and Stroock (1985), Karatzas and Shreve (1991), Stroock (2008), for instance.

Chapter 3
Malliavin Calculus

3.1 Cameron-Martin Theorem and Elementary IBP Formula

Let $(\mathcal{W}, \mathscr{B}(\mathcal{W}), \mu)$ be a probability space, where $\mathcal{W} = C_{(0)}([0, T]; \mathbb{R}^d)$, $\mathscr{B}(\mathcal{W})$ is the Borel σ-field and $\mu : \mathscr{B}(\mathcal{W}) \to [0, 1]$ is the Wiener measure.

For $a \in \mathcal{W}$ and $A \in \mathscr{B}(\mathcal{W})$, we define a set $A - a := \{\omega - a; \omega \in A\}$ and define a shifted measure $\mu(\cdot - a) : \mathscr{B}(\mathcal{W}) \ni A \mapsto \mu(A - a) \in [0, 1]$. Let $\mathcal{H}$ be the space given by

$$\mathcal{H} := \{h \in \mathcal{W}; \ h(\cdot) = \int_0^\cdot \dot{h}(s)ds, \ \dot{h} \in L^2([0, T]; \mathbb{R}^d)\} \subset \mathcal{W}.$$

We define $\langle h_1, h_2 \rangle_{\mathcal{H}} = \sum_{i=1}^d \int_0^T \dot{h}_1^i(t) \dot{h}_2^i(t) dt \ (= \langle \dot{h}_1, \dot{h}_2 \rangle_{L^2([0,T];\mathbb{R}^d)})$ for $h_1, h_2 \in \mathcal{H}$ and set $\|h\|_{\mathcal{H}} = \sqrt{\langle h, h \rangle_{\mathcal{H}}}$ for $h \in \mathcal{H}$. Then, $(\mathcal{H}, \langle \cdot, \cdot \rangle_{\mathcal{H}})$ becomes a Hilbert space. The dense subset $\mathcal{H}$ is called the Cameron-Martin space. For $h \in \mathcal{H}$, let $W(h)$ be the Wiener integral:

$$W(h) := \int_0^T \dot{h}(t) dW_t = \sum_{k=1}^d \int_0^T \dot{h}^k(t) dW_t^k \in L^2(\mathcal{W}).$$

The following theorem is important.

Theorem 3.16 (Cameron-Martin theorem) *Let $a \in \mathcal{W}$ be a shift. Then, the followings are equivalent:*

1. $\mu(\cdot - a)$ *is absolutely continuous with respect to μ,*
2. $a \in \mathcal{H}$.

If $h = \int_0^\cdot \dot{h}(s)ds \in \mathcal{H}$, we have

$$\frac{d\mu(\cdot - h)}{d\mu}(\omega) = \exp\left(W(h)(\omega) - \frac{1}{2}\|h\|_{\mathcal{H}}^2\right), \quad \omega \in \mathcal{W}.$$

The stochastic process W^h defined by $W_t^h(\omega) = W_t(\omega) - h(t) = \omega(t) - \int_0^t \dot{h}(s)ds$, $t \geq 0$ is a Brownian motion under the probability distribution $\mu(\cdot - h)$.

Proof See Theorem 1.3 with Exercise 1.5 in Shigekawa (2004). □

The Cameron-Martin theorem (Theorem 3.16) tells us that the Wiener measure is invariant under the shift for the direction of $\mathcal{H}$, also it is known that if $a \in \mathcal{W} \setminus \mathcal{H}$, the measure $\mu(\cdot - a)$ is singular with respect to μ. We now define the directional derivative for the direction of $\mathcal{H}$.

Let us define the following space:

$$\mathcal{S}(\mathcal{W}) := \Big\{ F : \mathcal{W} \to \mathbb{R}; \ F = f(W(h_1), \ldots, W(h_n)),$$
$$f \in C_b^\infty(\mathbb{R}^n), \ h_i \in \mathcal{H}, i = 1, \ldots, n, \ n \geq 1 \Big\}.$$

Note that $\mathcal{S}(\mathcal{W}) \overset{\text{dense}}{\subset} L^p(\mathcal{W})$ for all $p \geq 1$. For $F = f(W(h_1), \ldots, W(h_n)) \in \mathcal{S}(\mathcal{W})$, define

$$DF = \sum_{i=1}^n (\partial_i f)(W(h_1), \ldots, W(h_n)) h_i$$

or $\mathbb{R}^d$-valued stochastic process $\{D_t F\}_{t \geq 0} = \{(D_t^1 F, \ldots, D_t^d F)\}_{t \geq 0}$:

$$D_t F = \sum_{i=1}^n (\partial_i f)(W(h_1), \ldots, W(h_n)) \dot{h}_i(t), \quad 0 \leq t \leq T,$$

or

$$D_{\ell,t} F = \sum_{i=1}^n (\partial_i f)(W(h_1), \ldots, W(h_n)) \dot{h}_i^\ell(t), \quad 0 \leq t \leq T, \ \ell = 1, \ldots, d.$$

Here, $\langle DF, h \rangle_{\mathcal{H}}$ is regarded as the directional derivative of $F \in \mathcal{S}(\mathcal{W})$ for the direction $h \in \mathcal{H}$:

$$\langle DF(\omega), h \rangle_{\mathcal{H}} = \langle D.F(\omega), \dot{h}(\cdot) \rangle_{L^2([0,T]; \mathbb{R}^d)}$$
$$= \lim_{\epsilon \to 0} \frac{1}{\epsilon}[F(\omega + \epsilon h) - F(\omega)] = \frac{\partial}{\partial \epsilon} F(\omega + \epsilon h)|_{\epsilon = 0}, \quad \omega \in \mathcal{W}.$$

Indeed,

3.1 Cameron-Martin Theorem and Elementary IBP Formula

$$\frac{\partial}{\partial \epsilon} F(\omega + \epsilon h)|_{\epsilon=0} = \frac{\partial}{\partial \epsilon} f(W(h_1)(\omega + \epsilon h), \ldots, W(h_n)(\omega + \epsilon h))|_{\epsilon=0}$$

$$= \frac{\partial}{\partial \epsilon} f(\int_0^T \dot{h}_1(t) dW_t(\omega + \epsilon h), \ldots, \int_0^T \dot{h}_n(t) dW_t(\omega + \epsilon h))|_{\epsilon=0}$$

$$= \frac{\partial}{\partial \epsilon} f(\int_0^T \dot{h}_1(t) dW_t(\omega) + \epsilon \int_0^T \dot{h}_1(t) \dot{h}(t) dt, \ldots, \int_0^T \dot{h}_n(t) dW_t(\omega) + \epsilon \int_0^T \dot{h}_n(t) \dot{h}(t) dt)|_{\epsilon=0}$$

$$= \sum_{i=1}^n (\partial_i f)(W(h_1)(\omega), \ldots, W(h_n)(\omega)) \langle \dot{h}_i, \dot{h} \rangle_{L^2([0,T];\mathbb{R}^d)}$$

$$= \sum_{i=1}^n (\partial_i f)(W(h_1)(\omega), \ldots, W(h_n)(\omega)) \langle h_i, h \rangle_{\mathcal{H}} = \langle DF(\omega), h \rangle_{\mathcal{H}}, \quad \omega \in \mathcal{W}.$$

We have the following results.

Proposition 3.2 (Elementary integration by parts) *For $F \in \mathcal{S}(\mathcal{W})$ and $h \in \mathcal{H}$, we have*

$$\mathbb{E}[\langle DF, h \rangle_{\mathcal{H}}] = \mathbb{E}[FW(h)]. \tag{3.1}$$

Proof

$$\mathbb{E}[\langle DF, h \rangle_{\mathcal{H}}] = \int_{\mathcal{W}} \lim_{\epsilon \to 0} \frac{1}{\epsilon} \{F(\omega + \epsilon h) - F(\omega)\} d\mu(\omega)$$

$$= \lim_{\epsilon \to 0} \frac{1}{\epsilon} \{\int_{\mathcal{W}} F(\omega + \epsilon h) d\mu(\omega) - \int_{\mathcal{W}} F(\omega) d\mu(\omega)\}$$

$$= \lim_{\epsilon \to 0} \frac{1}{\epsilon} \int_{\mathcal{W}} F(\omega) \{e^{\epsilon W(h)(\omega) - \frac{\epsilon^2}{2} \|h\|_{\mathcal{H}}^2} - 1\} d\mu(\omega)$$

$$= \int_{\mathcal{W}} F(\omega) \lim_{\epsilon \to 0} \frac{1}{\epsilon} \{e^{\epsilon W(h)(\omega) - \frac{\epsilon^2}{2} \|h\|_{\mathcal{H}}^2} - 1\} d\mu(\omega)$$

$$= \int_{\mathcal{W}} F(\omega) W(h)(\omega) d\mu(\omega)$$

$$= \mathbb{E}[FW(h)].$$

$\square$

Proposition 3.3 (Elementary integration by parts II) *For $F, G \in \mathcal{S}(\mathcal{W})$ and $h \in \mathcal{H}$, we have*

$$\mathbb{E}[\langle DF, Gh \rangle_{\mathcal{H}}] = \mathbb{E}[F\{GW(h) - \langle DG, h \rangle_{\mathcal{H}}\}]. \tag{3.2}$$

Proof

$$\mathbb{E}[\langle DF, Gh\rangle_{\mathcal{H}}] = \mathbb{E}[\langle DF, h\rangle_{\mathcal{H}} G]$$
$$= \int_{\mathcal{W}} \lim_{\epsilon \to 0} \frac{1}{\epsilon}\{F(\omega+\epsilon h) - F(\omega)\}G(\omega)d\mu(\omega)$$
$$= \int_{\mathcal{W}} F(\omega) \lim_{\epsilon \to 0} \frac{1}{\epsilon}\{G(\omega-\epsilon h)e^{\epsilon W(h)(\omega) - \frac{\epsilon^2}{2}\|h\|_{\mathcal{H}}^2} - G(\omega)\}d\mu(\omega)$$
$$= \int_{\mathcal{W}} F(\omega)\{G(\omega)W(h)(\omega) - \langle DG, h\rangle_{\mathcal{H}}\}d\mu(\omega)$$
$$= \mathbb{E}[F\{GW(h) - \langle DG, h\rangle_{\mathcal{H}}\}].$$

□

3.2 Malliavin Derivative Operators and Sobolev Spaces

Using the elementary integration by parts in the previous subsection, we have the following, of which proof is given by Proposition 1.2.1 in Nualart (2006).

Proposition 3.4 *For* $p \geq 1$, $D : \mathcal{S}(\mathcal{W}) \overset{\text{dense}}{\subset} L^p(\mathcal{W}) \ni F \mapsto DF \in L^p(\mathcal{W}; \mathcal{H})$ *is a closable operator.*

For $F \in \mathcal{S}(\mathcal{W})$ and $k \in \mathbb{N}$, define

$$D^k F = \sum_{(i_1,\ldots,i_k)\in\{1,\ldots,n\}^k} (\partial_{i_1}\cdots\partial_{i_k}f)(W(h_1),\ldots,W(h_n))h_{i_1}\otimes\cdots\otimes h_{i_k}. \quad (3.3)$$

Generally, for $k \in \mathbb{N}$ and $p \geq 1$, $D^k : \mathcal{S}(\mathcal{W}) \overset{\text{dense}}{\subset} L^p(\mathcal{W}) \to L^p(\mathcal{W}; \mathcal{H}^{\otimes k})$ is a closable operator. For $k \in \mathbb{N}$ and $p \geq 1$, let $\|\cdot\|_{k,p}$ be given by

$$\|F\|_{k,p} = \|F\|_p + \sum_{j=1}^{k} \mathbb{E}[\|D^j F\|_{\mathcal{H}^{\otimes j}}^p]^{\frac{1}{p}}$$
$$= \sum_{j=0}^{k} \mathbb{E}[\sum_{\ell_1,\ldots,\ell_j=1}^{d} (\int_{0<t_1<\cdots<t_j<T} |D_{\ell_1,t_1}\cdots D_{\ell_j,t_j}F|^2 dt_1\cdots dt_j)^{p/2}]^{1/p}.$$

(3.4)

Definition 3.1 (*Sobolev space*) For $k \in \mathbb{N}$ and $p \geq 1$,

$$\mathbb{D}^{k,p} := \overline{\mathcal{S}(\mathcal{W})}^{\|\cdot\|_{k,p}}.$$

For $p \geq 1$, we define $\mathbb{D}^{0,p} := L^p(\mathcal{W})$. For $k \in \mathbb{N}$, we will denote the closed extension of the derivative operator $\bar{D}^k$ again by D^k called the Malliavin derivative operator.

3.3 Skorohod Integral

Definition 3.2 (*The space of smooth Wiener functionals in Malliavin sense*)

$$\mathbb{D}^\infty := \cap_{k \in \mathbb{N}} \cap_{p \geq 1} \mathbb{D}^{k,p}.$$

Theorem 3.17 *If $F \in (\mathbb{D}^{1,2})^N$ and $f \in C_b^1(\mathbb{R}^N)$, then $f(F) \in \mathbb{D}^{1,2}$ and it holds that*

$$Df(F) = \sum_{i=1}^{N} (\partial_i f)(F) DF^i.$$

Proof See the statement just before Proposition 1.2.3 in Nualart (2006). □

The above definitions can be extended to Hilbert-valued Wiener functionals. Let V be a real separable Hilbert space with the norm $\| \cdot \|_V := (\langle \cdot, \cdot \rangle_V)^{1/2}$. Let $\mathcal{S}_V(\mathcal{W}) := \{F : \mathcal{W} \to V; \ F = \sum_{j=1}^{n} F_j v_j, \ F_j \in \mathcal{S}(\mathcal{W}), \ v_j \in V, \ j \leq n, \ n \geq 1\}$ and define $D^k F = \sum_{j=1}^{n} D^k F_j \otimes v_j$, $k \geq 1$. Then, for $k \in \mathbb{N}$ and $p \geq 1$, $D^k : \mathcal{S}_V(\mathcal{W}) \overset{\text{dense}}{\subset} L^p(\mathcal{W}; V) \to L^p(\mathcal{W}; \mathcal{H}^{\otimes k} \otimes V))$ is a closable operator.

For $k \in \mathbb{N}$ and $p \geq 1$, let $\|F\|_{k,p,V} = \mathbb{E}[\|F\|_V^p]^{\frac{1}{p}} + \sum_{j=1}^{k} \mathbb{E}[\|D^j F\|_{\mathcal{H}^{\otimes j} \otimes V}^p]^{\frac{1}{p}}$ and $\mathbb{D}^{k,p}(V) := \overline{\mathcal{S}_V(\mathcal{W})}^{\| \cdot \|_{k,p,V}}$. For $p \geq 1$, let $\mathbb{D}^{0,p}(V) := L^p(V)$.

3.3 Skorohod Integral

We define the divergence operator δ as the adjoint of D.

Definition 3.3 (*The space of Skorohod integrable processes*)

$$\text{Dom}\delta := \{u. \in L^2([0,T] \times \mathcal{W}; \mathbb{R}^d); \ \exists c > 0 \ s.t.$$
$$|\mathbb{E}[\langle DF, u \rangle_{L^2([0,T])}]| \leq c \|F\|_2, \ \forall F \in \mathbb{D}^{1,2}\}.$$

For $u = (u^1, \ldots, u^d) \in \text{Dom}\delta$, there exists $\delta(u) = \sum_{i=1}^{d} \delta^i(u^i) \in L^2(\mathcal{W})$ such that

$$\mathbb{E}[\langle DF, u \rangle_{L^2([0,T])}] = \mathbb{E}[F \delta(u)] \qquad (3.5)$$

for all $F \in \mathbb{D}^{1,2}$, which is also represented by

$$\mathbb{E}[\int_0^T D_{i,t} F u_t^i dt] = \mathbb{E}[F \delta^i(u^i)], \quad i = 1, \ldots, d.$$

The above (3.5) is called the duality formula and $\delta(u)$ is called the *Skorohod integral* of u.

Theorem 3.18 *Let $G \in \mathbb{D}^{1,2}$, $u \in \text{Dom}\delta$ be such that $\mathbb{E}[G^2 \|u\|^2_{L^2([0,T])}]$, $\mathbb{E}[G^2\delta(u)^2]$ and $\mathbb{E}[\langle DG, u\rangle^2_{L^2([0,T])}]$ are finite. Then, $Fu \in \text{Dom}\delta$ and*

$$\delta(Gu) = G\delta(u) - \langle DG, u\rangle_{L^2([0,T])}. \tag{3.6}$$

Proof For $F \in \mathcal{S}(\mathcal{W})$, we have

$$\mathbb{E}[F\delta(Gu)] = E[\langle DF, Gu\rangle_\mathcal{H}] = \mathbb{E}[\langle G(DF), u\rangle_\mathcal{H}]$$
$$= \mathbb{E}[\langle D(FG) - F(DG), u\rangle_\mathcal{H}]$$
$$= \mathbb{E}[\langle D(FG), u\rangle_\mathcal{H}] - E[F\langle DG, u\rangle_\mathcal{H}]$$
$$= \mathbb{E}[F[G\delta(u) - \langle (DG), u\rangle_\mathcal{H}]]. \qquad \square$$

Let $\{\mathcal{F}_t\}$ be the Brownian filtration. If $u. \in L^2([0, T] \times \mathcal{W}; \mathbb{R}^d)$ is $\{\mathcal{F}_t\}_t$-adapted, then $u \in \text{Dom}\delta$ and $\delta(u)$ is given by the Itô integral:

$$\delta(u) = \int_0^T u_s dW_s = \sum_{k=1}^d \int_0^T u_s^k dW_s^k.$$

The following calculations hold.

Theorem 3.19 *For $u \in \mathbb{D}^{1,2}(H)$ be a $\{\mathcal{F}_t\}_t$-adapted process, we have*

$$D_{i,t}\int_0^T u_s dW_s^j = u_t \delta_{ij} + \int_t^T D_{i,t} u_s dW_s^j, \quad D_{i,t}\int_0^T u_s ds = \int_t^T D_{i,t} u_s ds.$$

Proof See (3.6) with Proposition 3.2.4 particularly, (3.4), and Example 3.4.4 with Proposition 3.4.3 in Nualart and Nualart (2018). $\qquad\square$

A functional belong to $L^2(\mathcal{W})$ has the following representation.

Theorem 3.20 (Itô's representation theorem) *For $F \in L^2(\mathcal{W})$, there exists $h \in L^2_a$ such that*

$$F = \mathbb{E}[F] + \int_0^T h_t dW_t. \tag{3.7}$$

Proof See Theorem 1.1.3 in Nualart (2006). $\qquad\square$

We give the representation of the adapted integrand if $F \in \mathbb{D}^{1,2}(\subset L^2(\mathcal{W}))$.

3.4 Malliavin's IBP Formula

Theorem 3.21 (Clark-Ocone formula) *For $F \in \mathbb{D}^{1,2}$, we have*

$$F = \mathbb{E}[F] + \int_0^T \mathbb{E}[D_t F | \mathcal{F}_t] dW_t. \tag{3.8}$$

Proof For any $Z \in L_a^2$,

$$\mathbb{E}[(\mathbb{E}[F] + \int_0^T \mathbb{E}[D_t F | \mathcal{F}_t] dW_t) \int_0^T Z_t dW_t]$$
$$= \mathbb{E}[\int_0^T \mathbb{E}[D_t F | \mathcal{F}_t] Z_t dt] = \int_0^T \mathbb{E}[\mathbb{E}[D_t F Z_t | \mathcal{F}_t]] dt$$
$$= \mathbb{E}[\int_0^T D_t F Z_t dt] = \mathbb{E}[F \int_0^T Z_t dW_t],$$

by the duality formula. □

More generally, for $k \in \mathbb{N}$, let $\text{Dom}\delta^k = \{u \in L^2(\mathcal{W}; \mathcal{H}^{\otimes k}); \exists C > 0 \text{ s.t. } |\mathbb{E}[\langle D^k F, u\rangle_{\mathcal{H}^{\otimes k}}]| \leq C \|F\|_{L^2(\mathcal{W})}, \forall F \in \mathbb{D}^{k,2}\}$. For $u \in \text{Dom}\delta^k$, there exists $\delta^k(u) \in L^2(\mathcal{W})$ such that

$$\mathbb{E}[\langle D^k F, u\rangle_{\mathcal{H}^{\otimes k}}] = \mathbb{E}[F \delta^k(u)]. \tag{3.9}$$

Related to Theorems 3.20 and 3.21, the following holds.

Theorem 3.22 (Stroock Taylor formula) *For $F \in \cap_{k \geq 1} \mathbb{D}^{k,2}$, we have*

$$F = \mathbb{E}[F] + \sum_{k \geq 1} \frac{1}{k!} \delta^k(\mathbb{E}[D^k F]). \tag{3.10}$$

Proof See Theorem 6, particularly, (8') in Stroock (1987). □

3.4 Malliavin's IBP Formula

Definition 3.4 (*Malliavin covariance matrix*) For $F = (F^1, \ldots, F^N) \in \mathbb{D}^\infty(\mathbb{R}^N)$, we define the Malliavin covariance matrix $\sigma^F = [\sigma_{ij}^F]_{1 \leq i,j \leq N}$:

$$\sigma_{ij}^F = \sum_{k=1}^d \int_0^T D_{k,t} F^i D_{k,t} F^j dt, \quad 1 \leq i, j \leq N.$$

Definition 3.5 (*Nondegenerate Wiener functional*) We say $F = (F^1, \ldots, F^N) \in \mathbb{D}^\infty(\mathbb{R}^N)$ is nondegenerate if σ^F is invertible a.s. and

$$\|\det(\sigma^F)^{-1}\|_p < \infty, \quad \forall p \geq 1.$$

Let $\mathcal{S}(\mathbb{R}^N)$ be the Schwartz space or the space of $\mathbb{R}$-valued rapidly decreasing functions on $\mathbb{R}^N$.

Theorem 3.23 (Malliavin's integration by parts) *Let $F = (F^1, \ldots, F^N) \in \mathbb{D}^\infty(\mathbb{R}^N)$ is nondegenerate, $G \in \mathbb{D}^\infty$ and $f \in \mathcal{S}(\mathbb{R}^N)$. Then, for any multi-index $\alpha \in \{1, \ldots, N\}^k$, there exists $H_\alpha(F, G) \in \mathbb{D}^\infty$ such that*

$$\mathbb{E}[\partial^\alpha f(F) G] = \mathbb{E}[f(F) H_\alpha(F, G)]. \tag{3.11}$$

Here, the Malliavin weight $H_\alpha(F, G)$ is given by

$$H_\alpha(F, G) = H_{(\alpha_k)}(F, H_{(\alpha_1, \ldots, \alpha_{k-1})}(F, G))$$

with

$$H_{(i)}(F, G) = \sum_{j=1}^N \delta(\gamma_{ij}^F DF^j G) = \sum_{j=1}^N \sum_{\ell=1}^d \delta^\ell(\gamma_{ij}^F D_{\ell, \cdot} F^j G), \quad i = 1, \ldots, N,$$

where γ^F is the inverse matrix of σ^F, i.e. $\gamma^F = (\sigma^F)^{-1}$, and has an upper bound: for $k \in \mathbb{N} \cup \{0\}$ and $p \geq 1$, there exist $q_1, q_2, q_3 > 1$ and $r \in \mathbb{N}$ such that

$$\|H_\alpha(F, G)\|_{k,p} \leq C \|\det(\sigma^F)^{-1}\|_{q_1}^r \|DF\|_{k+|\alpha|,q_2,\mathcal{H}}^{2Nr-|\alpha|} \|G\|_{k+|\alpha|,q_3}. \tag{3.12}$$

Proof Since $Df(F) = \sum_{i=1}^N \partial_i f(F) DF^i$, we have

$$\langle Df(F), DF^j \rangle_\mathcal{H} = \sum_{i=1}^N \partial_i f(F) \langle DF^i, DF^j \rangle_\mathcal{H}.$$

Then,

$$\mathbb{E}[\partial_i f(F) G] = \mathbb{E}[\langle Df(F), \sum_{j=1}^N \gamma_{ij}^F DF^j G \rangle_\mathcal{H}] = \mathbb{E}[f(F) \delta(\sum_{j=1}^N \gamma_{ij}^F DF^j G)].$$

Let $k \in \mathbb{N} \cup \{0\}$ and $p \geq 1$. We use a generic constant $C > 0$ depending on k, p whose value is varying from line to line. By Proposition 1.5.7 in Nualart (2006),

$$\|H_{(i)}(F, G)\|_{k,p}$$

3.4 Malliavin's IBP Formula

$$\leq \sum_{j=1}^{N} \|\delta([\sigma^F]_{ij}^{-1} DF^j G)\|_{k,p} \leq C \sum_{j=1}^{N} \|[\sigma^F]_{ij}^{-1} DF^j G\|_{k+1,p,\mathcal{H}}, \quad 1 \leq i \leq N.$$
(3.13)

Note that we have

$$D([\sigma^F]_{ij}^{-1} DF^j) = D[\sigma^F]_{ij}^{-1} DF^j + [\sigma^F]_{ij}^{-1} D^2 F^j$$

$$= -\sum_{k_1,k_2=1}^{N} [\sigma^F]_{ik_1}^{-1} [\sigma^F]_{k_2 j}^{-1} D[\sigma^F]_{k_1 k_2} DF^j + [\sigma^F]_{ij}^{-1} D^2 F^j \quad (3.14)$$

by the chain rule of Malliavin derivative and Lemma 2.1.6 of Nualart (2006). Then, using the Cramer formula $[\sigma^F]_{ij}^{-1} = \det(\sigma^F)^{-1}[\sigma^F]_{(ij)}$ where $[\sigma^F]_{(ij)}$ is the (i, j) minor matrix of the Malliavin covariance matrix σ^F, one can represent

$$D([\sigma^F]_{ij}^{-1} DF^j) = P_{2 \times 2(N-1)+1+2}(DF, D^2 F) \det(\sigma^F)^{-2}$$

$$= P_{4N-1}(DF, D^2 F) \det(\sigma^F)^{-2} \quad (3.15)$$

where $P_q(DF, D^2F, \ldots, D^r F)$ is a polynomial of the order q of $DF, D^2 F, \ldots, D^r F$. Iterating this operation, for $\ell \in \mathbb{N}$,

$$D^\ell([\sigma^F]_{ij}^{-1} DF^j) = P_{2(\ell+1)(N-1)+1+2\ell}(DF, D^2F, \ldots, D^{\ell+1}F) \det(\sigma^F)^{-(\ell+1)}$$

$$= P_{2N(\ell+1)-1}(DF, D^2F, \ldots, D^{\ell+1}F) \det(\sigma^F)^{-(\ell+1)} \quad (3.16)$$

as in the proof of Proposition 5.6 of Shigekawa (2004). Then, we have the upper bounds:

$$\|[\sigma^F]_{ij}^{-1} DF^j\|_{k+1,p,\mathcal{H}} \leq C \|\det(\sigma^F)^{-1}\|_{2(k+2)p}^{k+2} \|DF\|_{k+1,2(2N(k+2)-1)p,\mathcal{H}}^{2N(k+2)-1} \quad (3.17)$$

and

$$\|H_{(i)}(F, G)\|_{k,p} \leq C \|\det(\sigma^F)^{-1}\|_{2(k+2)p_1}^{k+2} \|DF\|_{k+1,2(2N(k+2)-1)p_1,\mathcal{H}}^{2N(k+2)-1} \|G\|_{k+1,p_2}$$
(3.18)

by using the Hölder inequality for Sobolev norms with $p_1, p_2 \geq 1$ such that $p_1^{-1} + p_2^{-1} = p^{-1}$ (see Proposition 1.5.6 of Nualart 2006). For $\alpha = (\alpha_1, \ldots, \alpha_\ell) \in \{1, \ldots, N\}^\ell$, we have

$$\|H_{(\alpha_1,\ldots,\alpha_\ell)}(F, G)\|_{k,p} = \|H_{(\alpha_\ell)}(F, H_{(\alpha_1,\ldots,\alpha_{\ell-1})}(F, G))\|_{k,p}$$
$$\leq C \|\det(\sigma^F)^{-1}\|_{2(k+2)p_1}^{k+2} \|DF\|_{k+1,2(2N(k+2)-1)p_1,\mathcal{H}}^{2N(k+2)-1} \|H_{(\alpha_1,\ldots,\alpha_{\ell-1})}(F, G)\|_{k+1,p_2}.$$
(3.19)

Iterating this procedure, for $k \in \mathbb{N} \cup \{0\}$ and $p \geq 1$, there exist $C > 0, q_1, q_2, q_3 > 1$ and $r \in \mathbb{N}$ such that

$$\|H_\alpha(F,G)\|_{k,p} \leq C \|\det(\sigma^F)^{-1}\|_{q_1}^r \|DF\|_{k+|\alpha|,q_2,\mathcal{H}}^{2Nr-|\alpha|} \|G\|_{k+|\alpha|,q_3}. \quad (3.20)$$

$\square$

As a consequence, the following important result is stated.

Theorem 3.24 (Malliavin's theorem) *If $F = (F^1, \ldots, F^N) \in \mathbb{D}^\infty(\mathbb{R}^N)$ is nondegenerate, F has a smooth density, i.e. $y \mapsto p^F(y) = \frac{d\mu \circ F^{-1}}{d\text{Leb}}(y)$ is smooth, where Leb is the Lebesgue measure.*

Proof See Theorem 4.9 in Malliavin and Thalmaier (2006). $\square$

3.5 Watanabe Distributions

Malliavin's theorem is extended by Watanabe's theory developed by S. Watanabe which also provides useful computational tools. Let $\mathcal{S}'(\mathbb{R}^N)$ be the dual of $\mathcal{S}(\mathbb{R}^N)$, i.e. $\mathcal{S}'(\mathbb{R}^N)$ is the space of Schwartz tempered distributions. Let $\mathbb{D}^{-\infty}$ be the dual space of $\mathbb{D}^\infty$, i.e. the space of continuous linear forms on $\mathbb{D}^\infty$.

Theorem 3.25 (Watanabe's theorem) *For a nondegenerate Wiener functional $F \in \mathbb{D}^\infty(\mathbb{R}^N)$, the map $\mathcal{S}(\mathbb{R}^N) \ni f \mapsto f \circ F \in \mathbb{D}^\infty$ is uniquely extended to the continuous linear map $\mathcal{S}'(\mathbb{R}^N) \ni T \mapsto T \circ F \in \mathbb{D}^{-\infty}$.*

Proof See Corollary of Theorem V.9.1 in Ikeda and Watanabe (1989). $\square$

Theorem 3.26 (Watanabe's duality) *For a nondegenerate Wiener functional $F \in \mathbb{D}^\infty(\mathbb{R}^N)$, $G \in \mathbb{D}^\infty$ and $T \in \mathcal{S}'(\mathbb{R}^N)$, it holds that*

$$_{\mathbb{D}^{-\infty}}\langle T(F), G \rangle_{\mathbb{D}^\infty} = {}_{\mathcal{S}'}\langle T, \mathbb{E}[G|F = \cdot] p^F(\cdot) \rangle_\mathcal{S} \quad (3.21)$$

where ${}_{\mathcal{S}'}\langle \cdot, \cdot \rangle_\mathcal{S}$ is the bilinear form on $\mathcal{S}'(\mathbb{R}^N)$ and $\mathcal{S}(\mathbb{R}^N)$, and ${}_{\mathbb{D}^{-\infty}}\langle T(F), G \rangle_{\mathbb{D}^\infty} = \mathbb{E}[T(F)G]$ is the pairing or the generalized expectation of $T(F) \in \mathbb{D}^{-\infty}$ and $G \in \mathbb{D}^\infty$.

Proof See Theorem 7.3 in Malliavin and Thalmaier (2006). $\square$

Therefore, for a nondegenerate Wiener functional $F \in \mathbb{D}^\infty(\mathbb{R}^N)$, the map $\mathcal{S}'(\mathbb{R}^N) \ni T \mapsto T(F) \in \mathbb{D}^{-\infty}$ is regarded as the adjoint of $u_F : \mathbb{D}^\infty \ni G \mapsto \mathbb{E}[G|F = \cdot] p^F(\cdot) \in \mathcal{S}(\mathbb{R}^N)$, i.e. $(u^F)^* : \mathcal{S}'(\mathbb{R}^N) \ni T \mapsto (u^F)^*(T) = T(F) \in \mathbb{D}^{-\infty}$.

3.5 Watanabe Distributions

$$
\begin{array}{cccc}
\mathbb{R}^N; & \mathcal{S}' & {}_{\mathcal{S}'}\langle T, \mathbb{E}[G|F=\cdot]p^F(\cdot)\rangle_{\mathcal{S}} & \mathcal{S} \\
\cdot \circ F = (u^F)^* \downarrow & & \| & \uparrow u^F = \mathbb{E}[\,\cdot\,|F=\,]p^F \\
\mathcal{W}; & \mathbb{D}^{-\infty} & {}_{\mathbb{D}^{-\infty}}\langle T(F), G\rangle_{\mathbb{D}^{\infty}} & \mathbb{D}^{\infty}
\end{array}
$$

We note that for any bounded measurable function $f : \mathbb{R}^N \to \mathbb{R}$, nondegenerate Wiener functional $F \in \mathbb{D}^\infty(\mathbb{R}^N)$ and smooth Wiener functional $G \in \mathbb{D}^\infty$, it holds that

$$\mathbb{E}[f(F)G] = \int_{\mathbb{R}^N} f(y)\,{}_{\mathbb{D}^{-\infty}}\langle \delta_y(F), G\rangle_{\mathbb{D}^\infty}\,dy. \tag{3.22}$$

In particular, we have the following representation of the density of nondegenerate Wiener functional.

Corollary 3.1 (Watanabe's representation) *For a nondegenerate Wiener functional $F \in \mathbb{D}^\infty(\mathbb{R}^N)$, we have ${}_{\mathbb{D}^{-\infty}}\langle \delta_y(F), 1\rangle_{\mathbb{D}^\infty} = {}_{\mathcal{S}'}\langle \delta_y, p^F\rangle_{\mathcal{S}} = p^F(y)$ for $y \in \mathbb{R}^d$.*

Proof See Theorem V.9.2 and its corollary in Ikeda and Watanabe (1989). □

Thus the density $p^F : y \mapsto \mathbb{E}[\delta_y(F)]$ is not only smooth but also in $\mathcal{S}(\mathbb{R}^N)$, i.e. a rapidly decreasing function. For $T \in \mathcal{S}'(\mathbb{R}^N)$, a multi-index $\alpha = (\alpha_1, \ldots, \alpha_k)$, a nondegenerate $F \in \mathbb{D}^\infty(\mathbb{R}^N)$ and $G \in \mathbb{D}^\infty$, we have

$${}_{\mathbb{D}^{-\infty}}\langle \partial^\alpha T(F), G\rangle_{\mathbb{D}^\infty} = {}_{\mathbb{D}^{-\infty}}\langle T(F), H_\alpha(F, G)\rangle_{\mathbb{D}^\infty}, \tag{3.23}$$

where $\partial^\alpha T = \partial_{\alpha_1} \cdots \partial_{\alpha_k} T$ is understood as the distributional derivative sense. We summarize the relationship of integration by parts computations for distributions in $\mathcal{S}'$ on the finite dimensional space $\mathbb{R}^N$ and those in $\mathbb{D}^{-\infty}$ on infinite dimensional space $\mathcal{W}$ as follows.

$$
\begin{array}{ccccc}
 & & {}_{\mathcal{S}'}\langle T, (-1)^{|\alpha|}\partial^\alpha(\mathbb{E}[G|F=\cdot]p^F(\cdot))\rangle_{\mathcal{S}} & & \\
 & & \| \qquad\qquad \| & & \\
\mathcal{S}' & {}_{\mathcal{S}'}\langle \partial^\alpha T, \mathbb{E}[G|F=\cdot]p^F(\cdot)\rangle_{\mathcal{S}} = {}_{\mathcal{S}'}\langle T, \mathbb{E}[H_\alpha(F, G)|F=\cdot]p^F(\cdot)\rangle_{\mathcal{S}} & \mathcal{S} \\
\downarrow \cdot \circ F & \| \qquad\qquad \| & \uparrow \mathbb{E}[\,\cdot\,|F=\,]p^F \\
\mathbb{D}^{-\infty} & {}_{\mathbb{D}^{-\infty}}\langle \partial^\alpha T(F), G\rangle_{\mathbb{D}^\infty} = {}_{\mathbb{D}^{-\infty}}\langle T(F), H_\alpha(F, G)\rangle_{\mathbb{D}^\infty} & \mathbb{D}^\infty
\end{array}
$$
$$\tag{3.24}$$

3.6 Malliavin Calculus for Multidimensional Diffusions

Consider the solution of the following stochastic differential equation:

$$dX_t^x = b(X_t^x) + \sum_{i=1}^{d} \sigma_i(X_t^x) dW_t^i, \quad X_0^x = x \in \mathbb{R}^N,$$

where $b \in C_b^\infty(\mathbb{R}^N; \mathbb{R}^N)$ and $\sigma := (\sigma_1, \ldots, \sigma_d) \in C_b^\infty(\mathbb{R}^N; \mathbb{R}^{N \times d})$.

Theorem 3.27 (Kunita's estimate) *For $t > 0$, the map $x \mapsto X_t^x$ is smooth, and for a multi-index $\alpha \in \{1, \ldots, N\}^k$ with $k \in \mathbb{N}$ and for $p \geq 1$, there exists $C > 0$ such that*

$$\sup_{t \in [0,T]} \left\| \frac{\partial^\alpha}{\partial x^\alpha} X_t^x \right\|_p \leq C, \quad \forall x \in \mathbb{R}^N. \tag{3.25}$$

For $x \in \mathbb{R}^N$, the Jacobian process $\{J_{0 \to t}\}_{t \geq 0}$ given by

$$J_{0 \to t} = \left[\frac{\partial}{\partial x_i} X_t^{x,j} \right]_{1 \leq i,j \leq N}, \quad t \geq 0$$

is invertible a.s.

Proof See Kunita (1990), Chapter V of Ikeda and Watanabe (1989) and Theorem 4.10.8 in Matsumoto and Taniguchi (2016). □

Theorem 3.28 (Malliavin differentiability and Malliavin derivative of solution to SDE) *For $t > 0$, $X_t^x \in \mathbb{D}^\infty(\mathbb{R}^N)$. In particular, we have*

$$D_s X_t^x = J_{0 \to t} J_{0 \to s}^{-1} \sigma(X_s^x), \quad s \leq t.$$

Proof of Theorem 3.28. By Theorem 3.19 and the chain rule of Malliavin derivative:

$$D_s X_t^x = \sigma(X_s^x) + \int_s^t \nabla \sigma(X_r^x) D_s X_r^x dW_r + \int_s^t \nabla b(X_r^x) D_s X_r^x dr.$$

Also, $J_{0 \to t}$ satisfies a linear SDE and we have

$$J_{0 \to t} J_{0 \to s}^{-1} = I + \int_s^t \nabla \sigma(X_r^x) J_{0 \to r} J_{0 \to s}^{-1} dW_r + \int_s^t \nabla b(X_r^x) J_{0 \to r} J_{0 \to s}^{-1} dr,$$

and

$$J_{0 \to t} J_{0 \to s}^{-1} \sigma(X_s^x) = \sigma(X_s^x) + \int_s^t \nabla \sigma(X_r^x) J_{0 \to r} J_{0 \to s}^{-1} \sigma(X_s^x) dW_r$$
$$+ \int_s^t \nabla b(X_r^x) J_{0 \to r} J_{0 \to s}^{-1} \sigma(X_s^x) dr.$$

3.6 Malliavin Calculus for Multidimensional Diffusions

Then, uniqueness of solution gives the assertion. □

Hereafter, we assume the uniformly elliptic condition, i.e. $\exists \epsilon > 0$ s.t.

$$\sum_{k=1}^{d} \sigma_k(x) \otimes \sigma_k(x) \geq \epsilon I_N, \quad \forall x \in \mathbb{R}^N, \tag{3.26}$$

or equivalently $\text{span}(\sigma_1(x), \ldots, \sigma_d(x)) = \mathbb{R}^N$ for all $x \in \mathbb{R}^N$.

Theorem 3.29 (Smoothness of density of solution to SDE) *For $t > 0$, we have*

$$\|(\det \sigma^{X_t^x})^{-1}\|_p < \infty, \quad \forall p \geq 1.$$

Then, the density $y \mapsto p^{X_t^x}(y) = E[\delta_y(X_t^x)]$ is in $\mathcal{S}(\mathbb{R}^d)$, i.e. is smooth and rapidly decreasing.

Proof Apply Theorem 4.9 in Malliavin and Thalmaier (2006) to the SDE under the current setting. □

In the elliptic case with $N = d$, a beautiful differentiation formula holds for the diffusion semigroup.

Theorem 3.30 (Bismut formula) *For $t > 0$,*

$$\frac{\partial}{\partial x}(P_t f)(x) = \mathbb{E}[f(X_t^x)\frac{1}{t}\int_0^t \sigma^{-1}(X_s^x) J_{0 \to s} dW_s].$$

Proof

$$\frac{\partial}{\partial x}(P_t f)(x) = \mathbb{E}[(\nabla f)(X_t^x) J_{0 \to t}]$$

$$= \frac{1}{t}\int_0^t \mathbb{E}[(\nabla f)(X_t^x) J_{0 \to t} J_{0 \to s}^{-1} \sigma(X_s^x) \sigma^{-1}(X_s^x) J_{0 \to s} ds]$$

$$= \frac{1}{t}\int_0^t \mathbb{E}[(\nabla f)(X_t^x) D_s X_t^x \sigma^{-1}(X_s^x) J_{0 \to s} ds] = \frac{1}{t}\mathbb{E}[\int_0^t D_s f(X_t^x) \sigma^{-1}(X_s^x) J_{0 \to s} ds]$$

$$= \frac{1}{t}\mathbb{E}[f(X_t^x) \int_0^t \sigma^{-1}(X_s^x) J_{0 \to s} dW_s].$$

□

We provide Kusuoka-Stroock's estimate for the inverse of Malliavin covariance matrix of elliptic Itô process and the integration by parts (Kusuoka and Stroock 1984), which is useful in our weak approximation analysis.

Theorem 3.31 (Kusuoka-Stroock estimate for elliptic Itô process) *Let $\psi : [0, T] \to [0, T]$ be a function such that $\psi(s) \leq s$ for all $s \in [0, T]$. Consider the following Itô process starting from $x \in \mathbb{R}^N$:*

$$\widetilde{X}_t^x = x + \int_0^t b(\widetilde{X}_{\psi(s)}^x) ds + \sum_{i=1}^d \int_0^t \sigma_i(\widetilde{X}_{\psi(s)}^x) dW_s^i, \quad t \geq 0,$$

where $b \in C_b^\infty(\mathbb{R}^N; \mathbb{R}^N)$ and $\sigma = (\sigma_1, \ldots, \sigma_d) \in C_b^\infty(\mathbb{R}^N; \mathbb{R}^{N \times d})$ with the condition (3.26), which is called the elliptic Itô process. Then, for $t > 0$, $\widetilde{X}_t^x \in \mathbb{D}^\infty(\mathbb{R}^N)$ is non-degenerate in the sense of Malliavin. In particular, for all $k \in \mathbb{N}$ and $p \geq 1$, there exists $c > 0$ such that for all $t > 0$ and $x \in \mathbb{R}^N$

$$\|D\widetilde{X}_t^x\|_{k,p,\mathcal{H}} \leq c t^{1/2}, \tag{3.27}$$

and for all $p \geq 1$, there exists $C > 0$ such that for all $t > 0$ and $x \in \mathbb{R}^N$

$$\|(\det \sigma^{\widetilde{X}_t^x})^{-1}\|_p \leq \frac{C}{t^N}. \tag{3.28}$$

Consequently, the following estimate holds: for a multi-index $\alpha \in \{1, \ldots, N\}^k$ with $k \in \mathbb{N}$ and $G \in \mathbb{D}^\infty$, there exist $C > 0$ and $p \geq 1$ such that

$$|\mathbb{E}[\partial^\alpha f(\widetilde{X}_t^x) G]| \leq C \|f\|_\infty \frac{1}{t^{|\alpha|/2}} \|G\|_{|\alpha|,p} \tag{3.29}$$

for all $f \in C_b^\infty(\mathbb{R}^N)$, $t > 0$ and $x \in \mathbb{R}^N$.

Proof See Theorems 2.19 and 3.5 of Kusuoka and Stroock (1984) for the proofs of (3.27) and (3.28). For the proof of (3.29), using (3.27) and (3.28) above and (3.12) of Theorem 20, we have: $\exists C_1, C_2 > 0, r, p, q > 1$ s.t.

$$\begin{aligned}|\mathbb{E}[\partial^\alpha f(\widetilde{X}_t^x)G]| &= |\mathbb{E}[f(\widetilde{X}_t^x) H_\alpha(\widetilde{X}_t^x, G)]| \\ &\leq \|f\|_\infty \|H_\alpha(\widetilde{X}_t^x, G)\|_1 \\ &\leq C_1 \|f\|_\infty \|(\det \sigma^{\widetilde{X}_t^x})^{-1}\|_q^r \|D\widetilde{X}_t^x\|_{|\alpha|,p,\mathcal{H}}^{2Nr-|\alpha|} \|G\|_{|\alpha|,p} \\ &\leq C_2 \|f\|_\infty \frac{1}{t^{|\alpha|/2}} \|G\|_{|\alpha|,p}. \quad \square\end{aligned}$$

For more details on Malliavin calculus, see Watanabe (1984), Ikeda and Watanabe (1989), Üstünel (1995), Malliavin (1997), Malliavin and Thalmaier (2006) and Nualart (2006).

Chapter 4
Asymptotic Expansion

We first review results on Watanabe's expansion on Wiener space. For $\{G_\varepsilon\}_{\varepsilon\in(0,1]} \subset \mathbb{D}^\infty$, we say $G_\varepsilon = O(\varepsilon^r)$ in $\mathbb{D}^\infty$ if for all $k \in \mathbb{N}$ and $p > 1$, $\|G_\varepsilon\|_{k,p} = O(\varepsilon^r)$ as $\varepsilon \downarrow 0$, i.e., $\limsup_{\varepsilon\downarrow 0} \frac{\|G_\varepsilon\|_{k,p}}{\varepsilon^r} < \infty$. Watanabe (1987) shows that if a family of Wiener functionals $\{F^\varepsilon\}_{\varepsilon\in(0,1]} \subset \mathbb{D}^\infty(\mathbb{R}^N)$ has an asymptotic expansion:

$$F^\varepsilon \sim F^0 + \varepsilon F_1 + \varepsilon^2 F_2 + \cdots \text{ in } \mathbb{D}^\infty(\mathbb{R}^N), \qquad (4.1)$$

where $F^0, F_1, F_2, \ldots \in \mathbb{D}^\infty(\mathbb{R}^N)$, in the sense that for any $m \geq 1$,

$$F^{\varepsilon,i} - (F^{0,i} + \varepsilon F_1^i + \varepsilon^2 F_2^i + \cdots + \varepsilon^m F_m^i) = O(\varepsilon^{m+1}) \text{ in } \mathbb{D}^\infty, \ i = 1,\ldots,N,$$

and satisfies the uniformly nondegenerate condition:

$$\limsup_{\varepsilon\downarrow 0} \|\det(\sigma^{F^\varepsilon})^{-1}\|_p < \infty \text{ for all } p > 1, \qquad (4.2)$$

then, for all $T \in \mathcal{S}'(\mathbb{R}^N)$, the composition $T(F^\varepsilon)$ is expanded in $\mathbb{D}^{-\infty}$. As a consequence, $\mathbb{E}[T(F^\varepsilon)]$ has an asymptotic expansion in $\mathbb{R}$:

$$\mathbb{E}[T(F^\varepsilon)] \sim a_0 + \varepsilon a_1 + \varepsilon^2 a_2 + \cdots, \qquad (4.3)$$

where

$$\begin{aligned} a_0 &= \mathbb{E}[T(F^0)], \ a_1 = \mathbb{E}[\textstyle\sum_{i=1}^N \partial_i T(F^0) F_1^i], \\ a_2 &= \mathbb{E}[\textstyle\sum_{i=1}^N \partial_i T(F^0) F_2^i] + \mathbb{E}[\tfrac{1}{2}\sum_{i_1,i_2=1}^N \partial_{i_1}\partial_{i_2} T(F^0) F_1^{i_1} F_1^{i_2}], \ \ldots \end{aligned} \qquad (4.4)$$

The method is called Watanabe's expansion which provide various applications since composition of irregular function as Schwartz distribution and diffusion process as

Wiener functional often appears in pure and applied mathematics, engineering and finance.

We give certain generalizations of the above asymptotic expansion in the chapter and will give an improvement as a weak approximation in the next chapter.

4.1 Asymptotic Expansion of Integrals of Wiener Functionals

First, we show an arbitrary m-order asymptotic expansion formula for expectation of composition of N-dimensional Wiener functional with a parameter ε and bounded measurable function $f : \mathbb{R}^N \to \mathbb{R}$. The uniform error bound of the form $const. \, \|f\|_\infty \varepsilon^{m+1}$ is shown and the uniformly nondegenerate condition of Watanabe (1987) is relaxed so that we are able to deal with wide practical applications.

Theorem 4.1 *Let $\{F^\varepsilon\}_{\varepsilon \in (0,1]} \subset \mathbb{D}^\infty(\mathbb{R}^N)$ be a family of Wiener functionals such that it has an asymptotic expansion in $\mathbb{D}^\infty(\mathbb{R}^N)$:*

$$F^\varepsilon \sim F^0 + \varepsilon F_1 + \varepsilon^2 F_2 + \cdots + \quad in \ \mathbb{D}^\infty(\mathbb{R}^N), \qquad (4.5)$$

and assume that

$$\|(\det \sigma^{F^0})^{-1}\|_p < \infty \qquad (4.6)$$

for all $p \in [1, \infty)$. Then, for $m \geq 1$, there exists $C > 0$ such that

$$\left| \mathbb{E}[f(F^\varepsilon)] - \left\{ \mathbb{E}[f(F^0)] + \sum_{j=1}^m \varepsilon^j \sum_{k,\alpha^{(k)},\beta^{(k)}}^{(j)} \mathbb{E}[f(F^0) H_{\alpha^{(k)}}(F^0, \prod_{e=1}^k F_{\beta_e}^{\alpha_e})] \right\} \right| \leq C \|f\|_\infty \varepsilon^{m+1} \qquad (4.7)$$

for any bounded measurable function $f : \mathbb{R}^N \to \mathbb{R}$ and $\varepsilon \in (0, 1]$, where

$$\sum_{k,\alpha^{(k)},\beta^{(k)}}^{(j)} = \sum_{k=1}^j \sum_{\substack{\beta^{(k)}=(\beta_1,\ldots,\beta_k)\in\mathbb{N}^k, \\ \sum_{\ell=1}^k \beta_\ell = j}} \sum_{\alpha^{(k)}=(\alpha_1,\ldots,\alpha_k)\in\{1,\ldots,N\}^k} \frac{1}{k!}. \qquad (4.8)$$

Proof of Theorem 4.1. We note that by (2.28) in Bally et al. (2016) there exists $C > 0$ such that

4.1 Asymptotic Expansion of Integrals of Wiener Functionals

$$\left|\det\sigma^{F^0+\lambda(F^\varepsilon-F^0)} - \det\sigma^{F^0}\right| \leq C\|D(F^\varepsilon-F^0)\|_{\mathcal{H}}(\|DF^0\|_{\mathcal{H}} + \|DF^\varepsilon\|_{\mathcal{H}})^{2N-1} \tag{4.9}$$

for all $\lambda \in [0, 1]$, and

$$\det\sigma^{F^0+\lambda(F^\varepsilon-F^0)} \geq \det\sigma^{F^0} - C\|D(F^\varepsilon-F^0)\|_{\mathcal{H}}(\|DF^0\|_{\mathcal{H}} + \|DF^\varepsilon\|_{\mathcal{H}})^{2N-1}. \tag{4.10}$$

Let $\psi \in C_b^\infty(\mathbb{R})$, $0 \leq \psi \leq 1$ be given by

$$\psi(x) = \mathbf{1}_{|x|\leq 1/4} + \exp\left(1 - (1/4)^2/((1/4)^2 - (x-1/4)^2)\right)\mathbf{1}_{1/4<|x|<1/2}, \quad x \in \mathbb{R}, \tag{4.11}$$

(e.g., (2.105) with $a = 1/4$ in Bally et al. (2016), and for $\varepsilon \in (0, 1]$, let

$$\eta^\varepsilon = \frac{C\|D(F^\varepsilon-F^0)\|_{\mathcal{H}}(\|DF^0\|_{\mathcal{H}} + \|DF^\varepsilon\|_{\mathcal{H}})^{2N-1}}{\det\sigma^{F^0}} \tag{4.12}$$

so that

$$\psi(\eta^\varepsilon) \neq 0 \quad \text{implies} \quad \det\sigma^{F^0+\lambda(F^\varepsilon-F^0)} \geq (1/2)\det\sigma^{F^0} \quad \text{for all } \lambda \in [0, 1]. \tag{4.13}$$

For $k \in \mathbb{N}$ and $p > 1$, there are $C, q > 0$ such that $\|\eta^\varepsilon\|_{k,p} \leq C\|F^\varepsilon - F^0\|_{k+1,q} = O(\varepsilon)$. By the properties of ψ, we have that for all $k \in \mathbb{N}$ and $p > 1$, $\|\psi(\eta^\varepsilon)\|_{k,p} = O(1)$,

$$\|1 - \psi(\eta^\varepsilon)\|_1 \leq \mathbb{P}(\eta^\varepsilon \geq 1/4) \leq 4^r \mathbb{E}[|\eta^\varepsilon|^r] = O(\varepsilon^r)$$

for arbitrary $r > 1$, as $1 - \psi(\eta^\varepsilon) \neq 0$ implies $\eta^\varepsilon \geq 1/4$, and furthermore for all $k \in \mathbb{N}$ and $p > 1$,

$$\|1 - \psi(\eta^\varepsilon)\|_{k,p} = O(\varepsilon^r)$$

for arbitrary $r > 1$, since for all $j \geq 1$, $\partial^j \psi(\eta^\varepsilon) \neq 0$ implies $1/4 < \eta^\varepsilon < 1/2$.

Let $f \in \mathcal{S}(\mathbb{R}^N)$ be a bounded function. Consider the decomposition

$$\mathbb{E}[f(F^\varepsilon)] = \mathbb{E}[f(F^\varepsilon)(1 - \psi(\eta^\varepsilon))] + \mathbb{E}[f(F^\varepsilon)\psi(\eta^\varepsilon)]. \tag{4.14}$$

For the first term of the right-hand side of (4.14), we have

$$|\mathbb{E}[f(F^\varepsilon)(1-\psi(\eta^\varepsilon))]| \leq O(\varepsilon^r)\|f\|_\infty,$$

for arbitrary $r > 1$. We next expand the second term of of the right-hand side of (4.14). We have

$$\mathbb{E}[f(F^\varepsilon)\psi(\eta^\varepsilon)]$$

$$= \mathbb{E}[f(F^0)\psi(\eta^\varepsilon)] + \sum_{i=1}^{m} \sum_{\alpha \in \{1,\ldots,N\}^i} \frac{1}{i!} \mathbb{E}[\partial^\alpha f(F^0) \prod_{\ell=1}^{i}(F^{\varepsilon,\alpha_\ell} - F^{0,\alpha_\ell})\psi(\eta^\varepsilon)] + R_f^1(\varepsilon)$$

$$= \mathbb{E}[f(F^0)] + \sum_{j=1}^{m} \varepsilon^j \sum_{k,\alpha,\beta}^{(j)} \mathbb{E}[\partial^\alpha f(F^0) \prod_{i=1}^{k} F_{\beta_i}^{\alpha_i}] + R_f^1(\varepsilon) + R_f^2(\varepsilon)$$

$$= \mathbb{E}[f(F^0)] + \sum_{j=1}^{m} \varepsilon^j \sum_{k,\alpha,\beta}^{(j)} \mathbb{E}[f(F^0) H_\alpha(F^0, \prod_{i=1}^{k} F_{\beta_i}^{\alpha_i})] + R_f^1(\varepsilon) + R_f^2(\varepsilon) \quad (4.15)$$

where $R_f^1(\varepsilon)$ is given by

$$R_f^1(\varepsilon) = \int_0^1 \frac{(1-\lambda)^m}{m!} \sum_{\alpha \in \{1,\ldots,N\}^{m+1}} \mathbb{E}[\partial^\alpha f(\widetilde{F^{\lambda,\varepsilon}}) \prod_{\ell=1}^{m+1}(F^{\varepsilon,\alpha_\ell} - F^{0,\alpha_\ell})\psi(\eta^\varepsilon)] d\lambda,$$

with $\widetilde{F^{\lambda,\varepsilon}} = F^0 + \lambda(F^\varepsilon - F^0)$, $\lambda \in [0,1]$, $\varepsilon \in (0,1]$, and $R_f^2(\varepsilon)$ has the form:

$$R_f^2(\varepsilon) = \sum_{\alpha \in \{1,\ldots,N\}^k, k \leq m} \mathbb{E}[\partial^\alpha f(F^0) G_\alpha^\varepsilon \psi(\eta^\varepsilon)]$$

$$+ \sum_{\alpha \in \{1,\ldots,N\}^k, k \leq m} \mathbb{E}[\partial^\alpha f(F^0) \hat{G}_\alpha^\varepsilon (1 - \psi(\eta^\varepsilon))]$$

with $\{G_\alpha^\varepsilon\}_{\alpha \in \{1,\ldots,N\}^k, k \leq m, \varepsilon \in (0,1]}$, $\{\hat{G}_\alpha^\varepsilon\}_{\alpha \in \{1,\ldots,N\}^k, k \leq m, \varepsilon \in (0,1]} \subset \mathbb{D}^\infty$ such that for all $k \leq m$ and $\alpha \in \{1,\ldots,d\}^k$, G_α^ε, $\hat{G}_\alpha^\varepsilon$, $\varepsilon \in (0,1]$ satisfy for all $\ell \in \mathbb{N}$, $p > 1$, $\|G_\alpha^\varepsilon\|_{\ell,p} = O(\varepsilon^{m+1})$, $\|\hat{G}_\alpha^\varepsilon\|_{\ell,p} = O(\varepsilon)$. By (4.13), we have

$$R_f^1(\varepsilon) = \int_0^1 \frac{(1-\lambda)^m}{m!} \sum_{\alpha \in \{1,\ldots,N\}^{m+1}} \mathbb{E}\left[f(\widetilde{F^{\lambda,\varepsilon}}) H_\alpha\left(\widetilde{F^{\lambda,\varepsilon}}, \prod_{\ell=1}^{m+1}(F^{\varepsilon,\alpha_\ell} - F^{0,\alpha_\ell})\psi(\eta^\varepsilon)\right)\right] d\lambda$$

with the estimate by (3.12): for $p \geq 1$, we have

$$\left\| H_\alpha\left(\widetilde{F^{\lambda,\varepsilon}}, \prod_{\ell=1}^{m+1}(F^{\varepsilon,\alpha_\ell} - F^{0,\alpha_\ell})\psi(\eta^\varepsilon)\right) \right\|_p$$

$$\leq C \|(\det \sigma^{F^0})^{-1}\|_{q_1}^r \|D\widetilde{F^{\lambda,\varepsilon}}\|_{|\alpha|,q_2,\mathcal{H}}^{2Nr-|\alpha|} \prod_{\ell=1}^{m+1} \|(F^{\varepsilon,\alpha_\ell} - F^{0,\alpha_\ell})\psi(\eta^\varepsilon)\|_{k_\ell,p_\ell}$$

for some $C > 0$, $r, q_1, q_2, p_1, \ldots, p_{m+1} \geq 1$ and $k_1, \ldots, k_{m+1} \in \mathbb{N}$. Here, we used the facts that $\psi(\eta^\varepsilon) \neq 0$ implies $(\det \sigma^{\widetilde{F^{\lambda,\varepsilon}}})^{-1} \leq 2(\det \sigma^{F^0})^{-1}$ with the assumption $(\det \sigma^{F^0})^{-1} \in \cap_{p \geq 1} L^p$, $\widetilde{F^{\lambda,\varepsilon}} \in (\mathbb{D}^\infty)^d$, and the estimates: for $k \in \mathbb{N}$ and $p >$

4.2 Small Noise Expansion

1, $\prod_{\ell=1}^{m+1}\|(F^{\varepsilon,\alpha_\ell} - F^{0,\alpha_\ell})\|_{k,p} = O(\varepsilon^{m+1})$, $\|\psi(\eta^\varepsilon)\|_{k,p} = O(1)$. Thus, there exists $C > 0$ such that

$$|R_f^1(\varepsilon)| \leq C\|f\|_\infty \varepsilon^{m+1}$$

for all $\varepsilon \in (0, 1]$. Also, we have the similar estimate for $R_f^2(\varepsilon)$, i.e. there exists $C > 0$ such that

$$|R_f^2(\varepsilon)| \leq \sum_{\alpha \in \{1,\ldots,N\}^k, k \leq m} |\mathbb{E}[f(F^0) H_\alpha(F^0, G_\alpha^\varepsilon \psi(\eta^\varepsilon))]|$$
$$+ \sum_{\alpha \in \{1,\ldots,N\}^k, k \leq m} |\mathbb{E}[f(F^0) H_\alpha(F^0, \hat{G}_\alpha^\varepsilon (1 - \psi(\eta^\varepsilon)))]|$$
$$\leq C\|f\|_\infty \varepsilon^{m+1}$$

for all $\varepsilon \in (0, 1]$, since we have for $\alpha \in \{1, \ldots, N\}^k$, $k \leq m$, for all $\ell \in \mathbb{N}$, $p > 1$, $\|G_\alpha^\varepsilon\|_{\ell,p} = O(\varepsilon^{m+1})$, and for all $\ell \in \mathbb{N}$, $p > 1$, $\|1 - \psi(\eta^\varepsilon)\|_{\ell,p} = O(\varepsilon^q)$ for arbitrary $q > 1$.

Therefore, (4.20) holds for $f \in \mathcal{S}'(\mathbb{R}^N)$, and thus for any bounded measurable function $f : \mathbb{R}^N \to \mathbb{R}$, we have

$$\mathbb{E}[f(F^\varepsilon)] = \mathbb{E}[f(F^0)] + \sum_{j=1}^m \varepsilon^{\nu_j} \sum_{k,\alpha,\beta}^{(j)} \mathbb{E}\Big[f(F^0) H_\alpha\Big(F^0, \prod_{i=1}^k F_{\beta_i}^{\alpha_i}\Big)\Big] + \widetilde{R}_f(\varepsilon)$$

with the estimate

$$|\widetilde{R}_f(\varepsilon)| \leq C\|f\|_\infty \varepsilon^{m+1}$$

for some $C > 0$ independent of f and ε. □

4.2 Small Noise Expansion

Consider the following small noise diffusion:

$$dX_t^{x,\varepsilon} = \sigma_0(X_t^{x,\varepsilon}) dt + \varepsilon \sum_{i=1}^d \sigma_i(X_t^{x,\varepsilon}) dW_t^i, \quad X_0^{x,\varepsilon} = x \in \mathbb{R}^N, \quad (4.16)$$

where $\sigma_i \in C_b^\infty(\mathbb{R}^N; \mathbb{R}^N)$, $i = 0, 1, \ldots, d$. Assume the uniformly elliptic condition. Note that we have

$$X_t^{x,\varepsilon} \sim X_t^{x,0} + \varepsilon X_{1,t}^x + \varepsilon^2 X_{2,t}^x + \cdots \text{ in } \mathbb{D}^\infty(\mathbb{R}^N) \quad (4.17)$$

where $X_t^{x,0}$ is the solution to $\frac{d}{dt}x_t^x = \sigma_0(x_t^x)$, $x_0^x = x$ and $X_{i,t}^x = \frac{1}{i!}\frac{\partial^i}{\partial \varepsilon^i}X_t^{x,\varepsilon}|_{\varepsilon=0}$, $i \in \mathbb{N}$. As $X_t^{x,0}$ degenerates, we define

$$F_t^{x,\varepsilon} := \frac{X_t^{x,\varepsilon} - X_t^{x,0}}{\varepsilon} \sim F_t^{x,0} + \varepsilon F_{1,t}^x + \cdots \text{ in } \mathbb{D}^\infty(\mathbb{R}^N), \qquad (4.18)$$

where $F_t^{x,0} = X_{1,t}^x$, $F_{1,t}^x = X_{2,t}^x$, …, so that $F_t^{x,0}$ is non-degenerate. Let $\bar{X}_t^{x,\varepsilon} = X_t^{x,0} + \varepsilon F_t^{x,0}$. We have the following.

Corollary 4.2 *For $m \geq 1$, there exists $C > 0$ such that*

$$\sup_{x \in \mathbb{R}^N} \left| \mathbb{E}[f(X_t^{x,\varepsilon})] - \left\{ \mathbb{E}[f(\bar{X}_t^{x,\varepsilon})] \right.\right.$$
$$\left.\left. + \sum_{j=1}^m \varepsilon^j \sum_{k,\alpha^{(k)},\beta^{(k)}}^{(j)} \mathbb{E}[f(\bar{X}_t^{x,\varepsilon})H_{\alpha^{(k)}}(F_t^{x,0}, \prod_{e=1}^k F_{\beta_e,t}^{x,\alpha_e})]\right\}\right|$$
$$\leq C\|f\|_\infty \varepsilon^{m+1} \qquad (4.19)$$

for any bounded measurable function $f : \mathbb{R}^N \to \mathbb{R}$ and $\varepsilon \in (0, 1]$.

Proof of Corollary 4.2. Let us define

$$\eta_t^{x,\varepsilon} = \frac{C\|D(F_t^{x,\varepsilon} - F_t^{x,0})\|_\mathcal{H}(\|DF_t^{x,0}\|_\mathcal{H} + \|DF_t^{x,\varepsilon}\|_\mathcal{H})^{2N-1}}{\det \sigma^{F_t^{x,0}}}$$

motivated by

$$\sup_{\lambda \in [0,1]} \left| \det \sigma^{F_t^{x,0}+\lambda(F_t^{x,\varepsilon}-F_t^{x,0})} - \det \sigma^{F_t^{x,0}} \right|$$
$$\leq C\|D(F_t^{x,\varepsilon} - F_t^{x,0})\|_\mathcal{H}(\|DF_t^{x,0}\|_\mathcal{H} + \|DF_t^{x,\varepsilon}\|_\mathcal{H})^{2N-1}.$$

Let $\psi(x) = \mathbf{1}_{|x| \leq 1/4} + \exp\left(1 - (1/4)^2/((1/4)^2 - (x-1/4)^2)\right)\mathbf{1}_{1/4 < |x| < 1/2}$, $x \in \mathbb{R}$. Consider the decomposition $\mathbb{E}[f(X_t^{x,\varepsilon})] = \mathbb{E}[f(X_t^{x,\varepsilon})\psi(\eta_t^{x,\varepsilon})] + \mathbb{E}[f(X_t^{x,\varepsilon})\{1 - \psi(\eta_t^{x,\varepsilon})\}]$. We have the expansion:

$$\mathbb{E}[f(X_t^{x,\varepsilon})\psi(\eta_t^{x,\varepsilon})]$$
$$= \int_{\mathbb{R}^N} f(X_t^{x,0} + \varepsilon y)\mathbb{E}[\delta_y(F_t^{x,0})\{1 + \sum_{j=1}^m \varepsilon^j \sum_{k,\alpha,\beta}^{(j)} H_\alpha(F_t^{x,0}, \prod_{i=1}^k F_{\beta_i,t}^{x,\alpha_i})\}]dy$$
$$+ R_f^1(t, x, \varepsilon) + R_f^2(t, x, \varepsilon), \qquad (4.20)$$

4.3 Small Time Expansion

where

$$R_f^1(t, x, \varepsilon) = \int_0^1 \frac{(1-\lambda)^m}{m!}$$

$$\sum_{\alpha \in \{1,\ldots,N\}^{m+1}} \mathbb{E}\Big[f(X_t^{x,0} + \varepsilon \widetilde{F_t^{x,\lambda,\varepsilon}}) H_\alpha\Big(\widetilde{F_t^{x,\lambda,\varepsilon}}, \prod_{\ell=1}^{m+1} (F_t^{x,\varepsilon,\alpha_\ell} - F_t^{x,0,\alpha_\ell}) \psi(\eta_t^\varepsilon)\Big)\Big] d\lambda$$

and

$$R_f^2(t, x, \varepsilon) = \sum_{\alpha \in \{1,\ldots,N\}^k, k \leq m} \mathbb{E}[f(X_t^{x,0} + \varepsilon F_t^{x,0}) H_\alpha(F_t^{x,0}, G_\alpha^{x,\varepsilon}(t) \psi(\eta_t^\varepsilon))]$$

$$+ \sum_{\alpha \in \{1,\ldots,N\}^k, k \leq m} \mathbb{E}[f(X_t^{x,0} + \varepsilon F_t^{x,0}) H_\alpha(F_t^{x,0}, \hat{G}_\alpha^{x,\varepsilon}(t)(1 - \psi(\eta_t^\varepsilon)))].$$

Then, the similar estimates as in the proof of Theorem 4.1 give the assertion. □

4.3 Small Time Expansion

We consider the following diffusion:

$$dX_t^x = \sigma_0(X_t^x)dt + \sum_{i=1}^d \sigma_i(X_t^x)dW_t^i, \quad X_0^x = x \in \mathbb{R}^N, \tag{4.21}$$

where $\sigma_i \in C_b^\infty(\mathbb{R}^N; \mathbb{R}^N), i = 0, 1, \ldots, d$. Assume the uniformly elliptic condition. Let us consider the following scaling diffusion with a small parameter $\varepsilon \in (0, 1]$:

$$dX_t^{x,\varepsilon} = \varepsilon^2 \sigma_0(X_t^{x,\varepsilon})dt + \varepsilon \sum_{i=1}^d \sigma_i(X_t^{x,\varepsilon})dW_t^i, \quad X_0^{x,\varepsilon} = x \in \mathbb{R}^N. \tag{4.22}$$

Since X_t^x and $X_1^{x,\sqrt{t}}$ has the same probability law, we analyze $X_1^{x,\varepsilon}$. We have

$$X_1^{x,\varepsilon} \sim x + \varepsilon F_0^x + \varepsilon^2 F_1^x + \cdots \text{ in } \mathbb{D}^\infty(\mathbb{R}^N) \tag{4.23}$$

with

$$F_0^x = \sum_{i=1}^d \sigma_i(x) W_1^i, \quad F_j^x = \sum_{\|\nu\|=j+1} \mathcal{L}_{\nu_1} \cdots \mathcal{L}_{\nu_{i-1}} \sigma_{\nu_i}(x) I_\nu(1), \quad j \in \mathbb{N} \tag{4.24}$$

where $\|\nu\| := 2\#\{i; \nu_i = 0\} + \#\{i; \nu_i \neq 0\}$,

$$\mathcal{L}_0 = \sum_{i=1}^{N} \sigma_0^i(\cdot)\partial_i + \frac{1}{2}\sum_{i_1,i_2=1}^{N}\sum_{j=1}^{d} \sigma_j^{i_1}(\cdot)\sigma_j^{i_2}(\cdot)\partial_{i_1}\partial_{i_2}, \quad \mathcal{L}_j = \sum_{i=1}^{N} \sigma_j^i(\cdot)\partial_i, \quad j=1,\ldots,d \tag{4.25}$$

and

$$I_\alpha(t) = \int_{0<t_1<\cdots<t_k<t} dW_{t_1}^{\alpha_1}\cdots dW_{t_k}^{\alpha_k}, \quad t>0, \ \alpha \in \{0,1,\ldots,d\}^k \tag{4.26}$$

with the notation $dW_t^0 = dt$. We define

$$F^{x,\varepsilon} := \frac{X_1^{x,\varepsilon} - x}{\varepsilon} \sim F_0^x + \varepsilon F_1^x + \cdots \quad \text{in } \mathbb{D}^\infty(\mathbb{R}^N). \tag{4.27}$$

Let $\bar{X}_1^{x,\varepsilon} = x + \varepsilon F_0^x = x + \varepsilon \sum_{i=1}^{d} \sigma_i(x) W_1^i$. We have the following small time expansion.

Corollary 4.3 *For $m \geq 1$, there exists $C > 0$ such that*

$$\sup_{x \in \mathbb{R}^N} \left| \mathbb{E}[f(X_t^x)] - \left\{ \mathbb{E}[f(\bar{X}_1^{x,\sqrt{t}})] \right.\right.$$
$$\left.\left. + \sum_{j=1}^{m} t^{j/2} \sum_{k,\alpha^{(k)},\beta^{(k)}} \mathbb{E}[f(\bar{X}_1^{x,\sqrt{t}}) H_{\alpha^{(k)}}(\sum_{\ell=1}^{d} \sigma_\ell(x) W_1^\ell, \prod_{e=1}^{k} F_{\beta_e}^{x,\alpha_e})] \right\} \right|$$
$$\leq C \|f\|_\infty t^{(m+1)/2} \tag{4.28}$$

for any bounded measurable function $f : \mathbb{R}^N \to \mathbb{R}$ and $t \in (0,1]$.

Proof of Corollary 4.3. Apply the similar argument of Corollary 4.2 for the solution of (4.22) and put $t=1$, $\varepsilon = \sqrt{t}$. □

4.4 Expansion Around One-Step Euler-Maruyama Scheme

Again consider the following diffusion:

$$dX_t^x = \sigma_0(X_t^x)dt + \sum_{i=1}^{d} \sigma_i(X_t^x)dW_t^i, \quad X_0^x = x \in \mathbb{R}^N, \tag{4.29}$$

where $\sigma_i \in C_b^\infty(\mathbb{R}^N; \mathbb{R}^N)$, $i=0,1,\ldots,d$ and assume the uniformly elliptic condition. Let $\bar{X}_t^{\text{EM},x}$ be the one-step Euler-Maruyama scheme, i.e. $\bar{X}_t^{\text{EM},x} = x + \sigma_0(x)t +$

4.4 Expansion Around One-Step Euler-Maruyama Scheme

$\sum_{i=1}^{d} \sigma_i(x) W_t^i$. We give an expansion without a perturbation parameter with a small time estimate and the bound of measurable test function.

Corollary 4.4 *For $m \geq 1$, there exists $C > 0$ such that*

$$\sup_{x \in \mathbb{R}^N} \left| \mathbb{E}[f(X_t^x)] - \left\{ \mathbb{E}[f(\bar{X}_t^{EM,x})] \right. \right.$$
$$\left. \left. + \sum_{j=1}^{m} \sum_{k, \alpha^{(k)}, \beta^{(k)}}^{(j)} \mathbb{E}[f(\bar{X}_t^{EM,x}) H_{\alpha^{(k)}}(\bar{X}_t^{EM,x}, \prod_{e=1}^{k} F_{\beta_e,t}^{x,\alpha_e})] \right\} \right|$$
$$\leq C \|f\|_\infty t^{(m+1)/2} \qquad (4.30)$$

for any bounded measurable function $f : \mathbb{R}^N \to \mathbb{R}$ and $t \in (0, 1]$, where

$$F_{j,t}^x = \sum_{|\nu|=j+1} \mathcal{L}_{\nu_1} \cdots \mathcal{L}_{\nu_j} \sigma_{\nu_{j+1}}(x) I_\nu(t), \quad j \in \mathbb{N}. \qquad (4.31)$$

Proof of Corollary 4.4. Let $\psi \in C_b^\infty(\mathbb{R})$, $0 \leq \psi \leq 1$ be given by

$$\psi(x) = \mathbf{1}_{|x| \leq 1/4} + \exp\left(1 - (1/4)^2 / ((1/4)^2 - (x - 1/4)^2)\right) \mathbf{1}_{1/4 < |x| < 1/2}, \quad x \in \mathbb{R},$$

and by considering $|\det \sigma^{X_t^x} - \det \sigma^{\bar{X}_t^{EM,x}}| \leq C \|D(X_t^x - \bar{X}_t^{EM,x})\|_{\mathcal{H}} (\|D\bar{X}_t^{EM,x}\|_{\mathcal{H}} + \|DX_t^x\|_{\mathcal{H}})^{2N-1}$, let

$$\eta_t = \frac{C \|D(X_t^x - \bar{X}_t^{EM,x})\|_{\mathcal{H}} (\|D\bar{X}_t^{EM,x}\|_{\mathcal{H}} + \|DX_t^x\|_{\mathcal{H}})^{2N-1}}{\det \sigma^{\bar{X}_t^{EM,x}}} \qquad (4.32)$$

so that

$$\psi(\eta_t) \neq 0 \quad \text{implies} \quad \det \sigma^{\bar{X}_t^{EM,x} + \lambda(X_t^x - \bar{X}_t^{EM,x})} \geq (1/2) \det \sigma^{\bar{X}_t^{EM,x}} \; \forall \lambda \in [0, 1]. \qquad (4.33)$$

Note that for $k \in \mathbb{N}$, $p > 1$, $\|X_t^x - \bar{X}_t^{EM,x}\|_{k,p} = O(t)$, $\|D\bar{X}_t^{EM,x}\|_{k,p,\mathcal{H}} = O(t^{1/2})$, $\|DX_t^x\|_{k,p,\mathcal{H}} = O(t^{1/2})$, $|(\det \sigma^{\bar{X}_t^{EM,x}})^{-1}| = O(t^{-N})$ ($\det \sigma^{\bar{X}_t^{EM,x}}$ is deterministic!). For $k \in \mathbb{N}$ and $p > 1$, $\|\eta_t\|_{k,p} = O(t^{1+(1/2)(2N-1)-N}) = O(t^{1/2})$ and $\|1 - \psi(\eta_t)\|_{k,p} = O(t^r)$ for arbitrary $r > 1$. For $f \in \mathcal{S}(\mathbb{R}^N)$,

$$\mathbb{E}[f(X_t^x) \psi(\eta_t)]$$
$$= \mathbb{E}[f(\bar{X}_t^{EM,x})] + \sum_{j=1}^{m} \varepsilon^j \sum_{k,\alpha,\beta}^{(j)} \mathbb{E}[f(\bar{X}_t^{EM,x}) H_\alpha(\bar{X}_t^{EM,x}, \prod_{i=1}^{k} F_{\beta_i,t}^{x,\alpha_i})]$$
$$+ R_f^1(t, x) + R_f^2(t, x) \qquad (4.34)$$

where $R_f^1(t, x)$ is given by

$$R_f^1(t, x) = \int_0^1 \frac{(1-\lambda)^m}{m!} \sum_{\alpha \in \{1,\ldots,N\}^{m+1}} \mathbb{E}\Big[f(\widetilde{X}_t^\lambda) H_\alpha\Big(\widetilde{X}_t^\lambda, \prod_{\ell=1}^{m+1}(X_t^{x,\alpha_\ell} - \bar{X}_t^{\mathrm{EM},x,\alpha_\ell})\psi(\eta_t)\Big)\Big] d\lambda$$

with $\widetilde{X}_t^\lambda = \bar{X}_t^{\mathrm{EM},x} + \lambda(X_t^x - \bar{X}_t^{\mathrm{EM},x})$, $\lambda \in [0, 1]$ and $R_f^2(t, x)$ has the form:

$$R_f^2(t, x) = \sum_{\alpha \in \{1,\ldots,N\}^k, k \leq m} \mathbb{E}[f(\bar{X}_t^{\mathrm{EM},x}) H_\alpha(\bar{X}_t^{\mathrm{EM},x}, G_\alpha(t)\psi(\eta_t))]$$
$$+ \sum_{\alpha \in \{1,\ldots,N\}^k, k \leq m} \mathbb{E}[f(\bar{X}_t^{\mathrm{EM},x}) H_\alpha(\bar{X}_t^{\mathrm{EM},x}, \hat{G}_\alpha(t)(1-\psi(\eta_t)))]$$

with $\{G_\alpha(t)\}_{\alpha \in \{1,\ldots,N\}^k, k \leq m, t \in (0,1]}$, $\{\hat{G}_\alpha(t)\}_{\alpha \in \{1,\ldots,N\}^k, k \leq m, t \in (0,1]} \subset \mathbb{D}^\infty$ such that for all $k \leq m$ and $\alpha \in \{1, \ldots, d\}^k$, $G_\alpha(t), \hat{G}_\alpha(t), t \in (0, 1]$ satisfy for all $\ell \in \mathbb{N}$, $p > 1$, $\|G_\alpha(t)\|_{\ell,p} = O(t^{(m+1)/2})$, $\|\hat{G}_\alpha(t)\|_{\ell,p} = O(t^{1/2})$. For $p \geq 1$, using (3.12), we have

$$\Big\|H_\alpha\Big(\widetilde{X}_t^\lambda, \prod_{\ell=1}^{m+1}(X_t^{x,\alpha_\ell} - \bar{X}_t^{\mathrm{EM},x,\alpha_\ell})\psi(\eta_t)\Big)\Big\|_p$$
$$\leq C \|(\det \sigma^{\bar{X}_t^{\mathrm{EM},x}})^{-1}\|_{q_1}^r \|D\widetilde{X}_t^\lambda\|_{|\alpha|,q_2,\mathcal{H}}^{2Nr-|\alpha|} \prod_{\ell=1}^{m+1} \|(X_t^{x,\alpha_\ell} - \bar{X}_t^{\mathrm{EM},x,\alpha_\ell})\psi(\eta_t)\|_{k_\ell, p_\ell}$$

for some $C > 0$, $r, q_1, q_2, p_1, \ldots, p_{m+1} \geq 1$ and $k, k_1, \ldots, k_{m+1} \in \mathbb{N}$, where we used the fact that $\psi(\eta_t) \neq 0$ implies $(\det \sigma^{\widetilde{X}_t^\lambda})^{-1} \leq 2(\det \sigma^{\bar{X}_t^{\mathrm{EM},x}})^{-1}$. Note that $(\det \sigma^{\bar{X}_t^{\mathrm{EM},x}})^{-1} = O(t^{-N})$ and for $k \in \mathbb{N}$ and $p > 1$, $\|D\widetilde{X}_t^\lambda\|_{k,p,\mathcal{H}} = O(t^{1/2})$, $\prod_{\ell=1}^{m+1} \|(X_t^{x,\alpha_\ell} - \bar{X}_t^{\mathrm{EM},x,\alpha_\ell})\|_{k,p} = O(t^{m+1})$, $\|\psi(\eta_t)\|_{k,p} = O(1)$. Thus, there exists $C > 0$ such that

$$\sup_{x \in \mathbb{R}^N} |R_f^1(t, x)| \leq C \|f\|_\infty t^{-rN} \sqrt{t}^{2rN-(m+1)} t^{m+1} = C \|f\|_\infty t^{(m+1)/2}$$

for all $t \in (0, 1]$. Using the properties: for $\alpha \in \{1, \ldots, N\}^k, k \leq m$, for all $\ell \in \mathbb{N}, p > 1$, $\|G_\alpha(t)\|_{\ell,p} = O(t^{(m+1)/2})$, and for all $\ell \in \mathbb{N}, p > 1$, $\|1 - \psi(\eta_t)\|_{\ell,p} = O(t^q)$ for arbitrary $q > 1$, i.e. there exists $C > 0$ such that

$$\sup_{x \in \mathbb{R}^N} |R_f^2(t, x)| \leq C \|f\|_\infty t^{(m+1)/2}$$

for all $t \in (0, 1]$. We obtain the assertion. $\square$

4.5 Explicit Computation and Generalization

We show an elementary example of the theoretical results of expansions.

Example 4.1 (*First-order expansion around one-step Euler-Maruyama scheme*) Consider a multidimensional elliptic diffusion driven by d-dimensional Brownian motion:

$$dX_t^x = \sigma_0(X_t^x)dt + \sum_{i=1}^{d} \sigma_i(X_t^x)dW_t^i, \quad X_0^x = x \in \mathbb{R}^N. \tag{4.35}$$

Then, the first order expansion is given by

$$\mathbb{E}[f(X_t^x)] = \mathbb{E}[f(\bar{X}_t^{\mathrm{EM},x})] + \mathbb{E}[f(\bar{X}_t^{\mathrm{EM},x}) \sum_{i=1}^{N} H_{(i)}(\bar{X}_t^{\mathrm{EM},x}, X_{t,1}^{x,i})] + \|f\|_\infty \times O(t),$$

where $X_{t,1}^{x,i} = \sum_{\alpha_1,\alpha_2=0}^{d} \mathcal{L}_{\alpha_1} \sigma_{\alpha_2}^i(x) I_{(\alpha_1,\alpha_2)}(t)$. In order to show its explicit representation, we prepare polynomials of Brownian motion inspired by Skorohod integral: for $k \in \mathbb{N}$ and $\alpha \in \{0, 1, \ldots, d\}^k$, $\mathbb{W}_{(\alpha_1)}(t) = W_t^{\alpha_1}$ if $k = 1$,

$$\mathbb{W}_{(\alpha_1,\ldots,\alpha_k)}(t) = W_t^{\alpha_k}\mathbb{W}_{(\alpha_1,\ldots,\alpha_{k-1})}(t) - \int_0^t D_s^{\alpha_k}\mathbb{W}_{(\alpha_1,\ldots,\alpha_{k-1})}(t)ds\mathbf{1}_{\alpha_k \neq 0} \quad \text{if } k \geq 2.$$

Lemma 4.3 *For $k \in \mathbb{N}$ and $\alpha \in \{0, 1, \ldots, d\}^k$, we have*

$$\mathbb{E}[\delta_y(W_t)I_\alpha(t)] = \mathbb{E}[\delta_y(W_t)\frac{1}{k!}\mathbb{W}_\alpha(t)].$$

Proof of Lemma 4.3. Let $\varphi \in C_b^\infty(\mathbb{R}^N)$. If $\alpha_k = 1, \ldots, d$,

$$\mathbb{E}[\delta_y(W_t) \int_0^t I_{(\alpha_1,\ldots,\alpha_{k-1})}(t_k)dW_{t_k}^{\alpha_k}]$$

$$= \mathbb{E}[\int_0^t D_{\alpha_k,t_k}\delta_y(W_t)I_{(\alpha_1,\ldots,\alpha_{k-1})}(t_k)dt_k]$$

$$= \sum_{i=1}^{d} \mathbb{E}[\int_0^t \partial_i \delta_y(W_t)D_{\alpha_k,t_k}W_t^i I_{(\alpha_1,\ldots,\alpha_{k-1})}(t_k)dt_k]$$

$$= \int_0^t \mathbb{E}[\partial_{\alpha_k}\delta_y(W_t) \int_0^{t_k} I_{(\alpha_1,\ldots,\alpha_{k-2})}(t_{k-1})dW_{t_{k-1}}^{\alpha_{k-1}}]dt_k,$$

if $\alpha_k = 0$,

$$\mathbb{E}[\delta_y(W_t) \int_0^t I_{(\alpha_1,\ldots,\alpha_{k-1})}(t_k)dW_{t_k}^{\alpha_k}] = \int_0^t \mathbb{E}[\delta_y(W_t) \int_0^{t_k} I_{(\alpha_1,\ldots,\alpha_{k-2})}(t_{k-1})dW_{t_{k-1}}^{\alpha_{k-1}}]dt_k.$$

Let α_* stand for a vector excluding zero elements from α. Then, iterating the procedure, we have

$$\mathbb{E}[\delta_y(W_t)I_\alpha(t)] = \mathbb{E}[\delta_y(W_t)\int_0^t I_{(\alpha_1,\ldots,\alpha_{k-1})}(t_k)dW_{t_k}^{\alpha_k}]$$

$$= \int_{0<t_1<\cdots<t_k<t} \mathbb{E}[\partial^{\alpha_*}\delta_y(W_t)]dt_1\cdots dt_k$$

$$= \mathbb{E}[\delta_y(W_t)\frac{1}{t^{|\alpha_*|}}\mathbb{W}_{\alpha_*}(t)]\frac{1}{k!}t^k = \mathbb{E}[\delta_y(W_t)\frac{1}{k!}\mathbb{W}_\alpha(t)]. \quad \square$$

The expansion term is calculated as follows:

$$\mathbb{E}\Big[f(\bar{X}_t^{\mathrm{EM},x})H_{(i)}\Big(\bar{X}_t^{\mathrm{EM},x},X_{t,1}^{x,i}\Big)\Big]$$

$$= \sum_{i=1}^N \mathbb{E}\Big[(\partial_i f)(\bar{X}_t^{\mathrm{EM},x})X_{t,1}^{x,i}\Big]$$

$$= \sum_{i=1}^N \sum_{\alpha_1,\alpha_2=0}^d \mathcal{L}_{\alpha_1}\sigma_{\alpha_2}^i(x)\mathbb{E}\Big[(\partial_i f)(\bar{X}_t^{\mathrm{EM},x})I_{(\alpha_1,\alpha_2)}(t)\Big]$$

$$= \sum_{i=1}^N \sum_{\alpha_1,\alpha_2=0}^d \mathcal{L}_{\alpha_1}\sigma_{\alpha_2}^i(x)\int_{\mathbb{R}^d}(\partial_i f)(x+\sigma_0(x)t+\sigma(x)y)\mathbb{E}[\delta_y(W_t)I_{(\alpha_1,\alpha_2)}(t)]dy$$

$$= \sum_{i=1}^N \sum_{\alpha_1,\alpha_2=0}^d \mathcal{L}_{\alpha_1}\sigma_{\alpha_2}^i(x)\int_{\mathbb{R}^d}(\partial_i f)(x+\sigma_0(x)t+\sigma(x)y)\mathbb{E}[\delta_y(W_t)\frac{1}{2}\mathbb{W}_{(\alpha_1,\alpha_2)}(t)]dy$$

$$= \sum_{i=1}^N \sum_{\alpha_1,\alpha_2=0}^d \mathcal{L}_{\alpha_1}\sigma_{\alpha_2}^i(x)\mathbb{E}[(\partial_i f)(\bar{X}_t^{\mathrm{EM},x})\frac{1}{2}\mathbb{W}_{(\alpha_1,\alpha_2)}(t)]$$

$$= \sum_{i=1}^N \sum_{\alpha_1,\alpha_2=0}^d \mathcal{L}_{\alpha_1}\sigma_{\alpha_2}^i(x)\mathbb{E}[f(\bar{X}_t^{\mathrm{EM},x})H_{(i)}(\bar{X}_t^{\mathrm{EM},x},\frac{1}{2}\mathbb{W}_{(\alpha_1,\alpha_2)}(t))]. \quad (4.36)$$

Here, the weight is explicitly obtained by

$$H_{(i)}(\bar{X}_t^{\mathrm{EM},x},\frac{1}{2}\mathbb{W}_{(\alpha_1,\alpha_2)}(t)) = \delta\Big(\sum_{j=1}^N \gamma_{ij}^{\bar{X}_t^{\mathrm{EM},x}} D\bar{X}_t^{x,j}\frac{1}{2}\mathbb{W}_{(\alpha_1,\alpha_2)}(t)\Big)$$

$$= \sum_{j=1}^N \sum_{\alpha_3=1}^d \frac{1}{2t}A_{ij}(x)\sigma_{\alpha_3}^j(x)\mathbb{W}_{(\alpha_1,\alpha_2,\alpha_3)}(t), \quad (4.37)$$

4.5 Explicit Computation and Generalization

where $A = (A_{i,j}(\cdot))_{1 \leq i,j \leq N}$ is a deterministic function such that for each $x \in \mathbb{R}^N$, $A_{i,j}(x)$ is the (i, j)-element of the inverse matrix of $(\sigma_{ij}^{\bar{X}_t^{\text{EM},x}})_{1 \leq i,j \leq N}$ with $\sigma_{ij}^{\bar{X}_t^{\text{EM},x}} = \sum_{k=1}^d \sigma_k^i(x)\sigma_k^j(x)$.

Then, we have

$$\mathbb{E}[f(X_t^x)]$$
$$= \mathbb{E}[f(\bar{X}_t^{\text{EM},x})]$$
$$+ \sum_{i,j=1}^N \sum_{\alpha_1,\alpha_2=0}^d \sum_{\alpha_1=1}^d \mathcal{L}_{\alpha_1}\sigma_{\alpha_2}^i(x)\frac{1}{2t}A_{ij}(x)\sigma_{\alpha_3}^j(x)\mathbb{E}[f(\bar{X}_t^{\text{EM},x})\mathbb{W}_{(\alpha_1,\alpha_2,\alpha_3)}(t)]$$
$$+ \|f\|_\infty \times O(t),$$

where $\mathbb{W}_{(\alpha_1,\alpha_2,\alpha_3)}(t)$ is the third-order polynomial:

$$\mathbb{W}_{(\alpha_1,\alpha_2,\alpha_3)}(t) = W_t^{\alpha_1} W_t^{\alpha_2} W_t^{\alpha_3} - W_t^{\alpha_1} t \mathbf{1}_{\alpha_2=\alpha_3 \neq 0} - W_t^{\alpha_2} t \mathbf{1}_{\alpha_1=\alpha_3 \neq 0} - W_t^{\alpha_3} t \mathbf{1}_{\alpha_1=\alpha_2 \neq 0}.$$

The above example suggests that in the expansion coefficients of form $\mathbb{E}[\partial^\alpha f(F^0) G]$ with $f \in \mathcal{S}'(\mathbb{R}^N)$, if G can be written by a divergence $G = \delta(u)$ of a stochastic process u, each coefficient may be obtained in an expectation with a polynomial of F^0.

In order to give a general representation of expansion coefficients, we show a "polynomialization" of Watanabe's expansion. The readers particularly interested in weak approximation may go to this chapter for their first reading.

Theorem 4.2 *Let $\{F^\varepsilon\}_{\varepsilon \in (0,1]} \subset \mathbb{D}^\infty(\mathbb{R}^N)$ be a family of Wiener functionals such that it has an asymptotic expansion in $\mathbb{D}^\infty(\mathbb{R}^N)$:*

$$F^\varepsilon \sim F^0 + \varepsilon F_1 + \varepsilon^2 F_2 + \cdots + \quad \text{in } \mathbb{D}^\infty(\mathbb{R}^N), \tag{4.38}$$

and assume that

$$\|(\det \sigma^{F^0})^{-1}\|_p < \infty \tag{4.39}$$

for all $p \in [1, \infty)$. Then, for $m \geq 1$, there exists $C > 0$ such that

$$\left| \mathbb{E}[f(F^\varepsilon)] - \left\{ \mathbb{E}[f(F^0)] + \sum_{j=1}^m \varepsilon^j \sum_{k, \alpha^{(k)}, \beta^{(k)}, \gamma^{(p)}}^{(j)} \right. \right.$$
$$\left. \left. \mathbb{E}\left[f(F^0) H_{\alpha^{(k)} * \gamma^{(p)}}\left(F^0, \frac{1}{p!} \langle DF^{0,\gamma_1} \otimes \cdots \otimes DF^{0,\gamma_p}, \mathbb{E}[D^p \prod_{e=1}^k F_{\beta_e}^{\alpha_e}] \rangle_{\mathcal{H}^{\otimes p}} \right) \right] \right\} \right|$$
$$\leq C\|f\|_\infty \varepsilon^{m+1}, \tag{4.40}$$

for any bounded measurable function $f : \mathbb{R}^N \to \mathbb{R}$ and $\varepsilon \in (0, 1]$, where

$$\sum_{k,\alpha^{(k)},\beta^{(k)},\gamma^{(p)}}^{(j)} = \sum_{k=1}^{j} \sum_{\substack{\beta^{(k)}=(\beta_1,\dots,\beta_k)\in\mathbb{N}^k \\ \sum_{\ell=1}^{k}\beta_\ell=j}} \sum_{\alpha^{(k)}=(\alpha_1,\dots,\alpha_k)\in\{1,\dots,N\}^k} \frac{1}{k!} \sum_{\gamma^{(p)}=(\gamma_1,\dots,\gamma_p)\in\{1,\dots,N\}^p,\, p\geq 0}. \tag{4.41}$$

Here, $\alpha^{(k)} * \gamma^{(p)}$ represents $\alpha^{(k)} * \gamma^{(p)} = (\alpha_1, \dots, \alpha_k, \gamma_1, \dots, \gamma_p)$ and we used the convention: if $p = 0$, $\frac{1}{p!}\langle DF^{0,\gamma_1} \otimes \cdots \otimes DF^{0,\gamma_p}, \mathbb{E}[D^p G]\rangle_{\mathcal{H}^{\otimes p}} = \mathbb{E}[G]$.

Proof Apply Theorem 4.1 with the following: for $f \in \mathcal{S}(\mathbb{R}^N)$ and $G = \prod_{i=1}^{k} F_{\beta_i}^{\alpha_i}$,

$\mathbb{E}[\partial^\alpha f(F^0)G]$
$= \mathbb{E}[\partial^\alpha f(F^0)]\mathbb{E}[G] + \mathbb{E}[\partial^\alpha f(F^0) \sum_{p\geq 1} \frac{1}{p!} \delta^p(\mathbb{E}[D^p G])]$
$= \mathbb{E}[\partial^\alpha f(F^0)]\mathbb{E}[G] + \sum_{p\geq 1} \frac{1}{p!} \mathbb{E}[\langle D^p \partial^\alpha f(F^0), \mathbb{E}[D^p G]\rangle_{\mathcal{H}^{\otimes p}}]$
$= \mathbb{E}[\partial^\alpha f(F^0)]\mathbb{E}[G] + \sum_{p\geq 1} \sum_{\gamma} \mathbb{E}[\partial^{\alpha*\gamma} f(F^0) \frac{1}{p!} \langle DF^{0,\gamma_1} \otimes \cdots \otimes DF^{0,\gamma_p}, \mathbb{E}[D^p G]\rangle_{\mathcal{H}^{\otimes p}}]$
$= \sum_{p\geq 0} \sum_{\gamma} \mathbb{E}[f(F^0) H_{\alpha*\gamma}(F^0, \frac{1}{p!} \langle DF^{0,\gamma_1} \otimes \cdots \otimes DF^{0,\gamma_p}, \mathbb{E}[D^p G]\rangle_{\mathcal{H}^{\otimes p}})]$,

where we used the Stroock-Taylor formula, the duality formula, the chain rule of Malliavin derivative and the IBP formula. $\square$

Remark 4.4 If F^0 is a zero mean Gaussian random variable, we have

$\mathbb{E}[\delta^p(\mathbb{E}[D^p G])|F^0 = y]p^{F^0}(y)$
$=_{\mathcal{S}'}\langle \delta_y(\cdot), \mathbb{E}[\delta^p(\mathbb{E}[D^p G])|F^0 = \cdot]p^{F^0}(\cdot)\rangle_{\mathcal{S}}$
$=_{\mathbb{D}^{-\infty}}\langle \delta_y(F^0), \delta^p(\mathbb{E}[D^p G])\rangle_{\mathbb{D}^\infty}$
$= \mathbb{E}[\delta_y(F^0)\delta^p(\mathbb{E}[D^p G])]$
$= \mathbb{E}[\sum_\gamma \partial^\gamma \delta_y(F^0)\langle DF^{0,\gamma_1} \otimes \cdots \otimes DF^{0,\gamma_p}, \mathbb{E}[D^p G]\rangle_{\mathcal{H}^{\otimes p}}]$
$= \sum_\gamma \mathbb{E}[\delta_y(F^0) H_\gamma(F^0, 1)]\langle DF^{0,\gamma_1} \otimes \cdots \otimes DF^{0,\gamma_p}, \mathbb{E}[D^p G]\rangle_{\mathcal{H}^{\otimes p}}$
$= \sum_\gamma {}_{\mathcal{S}'}\langle \delta_y(\cdot), \mathbb{E}[H_\gamma(F^0, 1)|F^0 = \cdot]p^{F^0}(\cdot)\rangle_{\mathcal{S}} DF^{0,\gamma_1} \otimes \cdots \otimes DF^{0,\gamma_p}, \mathbb{E}[D^p G]\rangle_{\mathcal{H}^{\otimes p}}$
$= \sum_\gamma \mathbb{E}[H_\gamma(F^0, 1)|F^0 = y]p^{F^0}(y)\langle DF^{0,\gamma_1} \otimes \cdots \otimes DF^{0,\gamma_p}, \mathbb{E}[D^p G]\rangle_{\mathcal{H}^{\otimes p}}$,

i.e.

$\mathbb{E}[\delta^p(\mathbb{E}[D^p G])|F^0 = y]$
$= \sum_\gamma \mathbb{E}[H_\gamma(F^0, 1)|F^0 = y]\langle DF^{0,\gamma_1} \otimes \cdots \otimes DF^{0,\gamma_p}, \mathbb{E}[D^p G]\rangle_{\mathcal{H}^{\otimes p}}$.

4.5 Explicit Computation and Generalization

Moreover, by (3.24), it holds that

$$\mathbb{E}[\delta^p(\mathbb{E}[D^p G])|F^0 = y] p^{F^0}(y)$$
$$= \sum_\gamma (-1)^{|\gamma|} \partial^\gamma p^{F^0}(y) \langle DF^{0,\gamma_1} \otimes \cdots \otimes DF^{0,\gamma_p}, \mathbb{E}[D^p G] \rangle_{\mathcal{H}^{\otimes p}}$$

or

$$\mathbb{E}[\delta^p(\mathbb{E}[D^p G])|F^0 = y] = \sum_\gamma \partial^\nu_\gamma(1) \langle DF^{0,\gamma_1} \otimes \cdots \otimes DF^{0,\gamma_p}, \mathbb{E}[D^p G] \rangle_{\mathcal{H}^{\otimes p}}$$

where $\partial^\nu_\gamma = \partial^\nu_{\gamma_1} \circ \cdots \circ \partial^\nu_{\gamma_p}$ for $\gamma = (\gamma_1, \ldots, \gamma_p) \in \{1, \ldots, N\}^p$ is the divergence with respect to the law $\nu = p^{F^0}(x)dx$ given by

$$\partial^\nu_\ell h(x) = (-\frac{\partial}{\partial x_\ell} \log p^{F^0}(x)) h(x) - \frac{\partial}{\partial x_\ell} h(x), \quad x \in \mathbb{R}^N, \ \ell = 1, \ldots, N.$$

See Chapter III.2.4 and III.3 of Malliavin (1997) for computation of divergence.

Remark 4.5 We can generalize the expansion for the fractional order case as follows. On an abstract Wiener space, let $\{F^\varepsilon\}_{\varepsilon \in (0,1]} \subset \mathbb{D}^\infty(\mathbb{R}^N)$ be a family of Wiener functionals such that F^ε has an asymptotic expansion in $\mathbb{D}^\infty(\mathbb{R}^N)$:

$$F^{\varepsilon,i} \sim F^{0,i} + \varepsilon^{\kappa_1} F_1^i + \varepsilon^{\kappa_2} F_2^i + \cdots \text{ in } \mathbb{D}^\infty, \ i = 1, \ldots, N, \tag{4.42}$$

where $F^0, F_1, F_2, \ldots \in \mathbb{D}^\infty(\mathbb{R}^N)$ and $\{\kappa_i; i \in \mathbb{N}\}$ satisfies $0 < \kappa_1 < \kappa_2 < \cdots$, in the sense that for any $m \geq 1$,

$$F^{\varepsilon,i} - (F^{0,i} + \varepsilon^{\kappa_1} F_1^i + \varepsilon^{\kappa_2} F_2^i + \cdots + \varepsilon^{\kappa_m} F_m^i) = O(\varepsilon^{\kappa_{m+1}}) \text{ in } \mathbb{D}^\infty, \ i = 1, \ldots, N, \tag{4.43}$$

and assume that the Malliavin covariance matrix σ^{F^0} is invertible a.s. and

$$\|(\det \sigma^{F^0})^{-1}\|_p < \infty \tag{4.44}$$

for all $p \geq 1$. Then, for $m \geq 1$, there exists $C > 0$ such that

$$\left| \mathbb{E}[f(F^\varepsilon)] - \left\{ \mathbb{E}[f(F^0)] + \sum_{j=1}^m \varepsilon^{\nu_j} \sum_{k,\alpha,\beta,\gamma}^{(j)} \mathbb{E}\left[f(F^0) H_{\alpha * \gamma}\left(F^0, \frac{1}{p!} \langle DF^{0,\gamma_1} \otimes \cdots \otimes DF^{0,\gamma_p}, \mathbb{E}[D^p \prod_{i=1}^k F_{\beta_i}^{\alpha_i 1}]\rangle_{\mathcal{H}^{\otimes p}} \right) \right] \right\} \right|$$
$$\leq C \|f\|_\infty \varepsilon^{\nu_{m+1}}, \tag{4.45}$$

for any bounded measurable function $f : \mathbb{R}^N \to \mathbb{R}$ and $\varepsilon \in (0, 1]$, where $\nu_\ell, \ell \in \mathbb{N}$ are all the elements of $\{\sum_{i=1}^m \kappa_{\beta_i}; \ \beta_1, \ldots, \beta_m \in \mathbb{N}, m \in \mathbb{N}\}$ in increasing order, and

$$\sum_{k,\alpha,\beta,\gamma}^{(j)} = \sum_{\substack{\beta=(\beta_1,\ldots,\beta_k)\in\mathbb{N}^k, k\in\mathbb{N},\\ \sum_{\ell=1}^{k}\kappa_{\beta_\ell}=\nu_j}} \sum_{\alpha=(\alpha_1,\ldots,\alpha_k)\in\{1,\ldots,N\}^k} \frac{1}{k!} \sum_{\gamma\in\{1,\ldots,N\}^p, p\geq 0} . \quad (4.46)$$

See Takahashi and Yamada (2024) for more details.

For the elliptic diffusion:

$$dX_t^{x,\varepsilon} = \sigma_0(X_t^{x,\varepsilon})dt + \varepsilon \sum_{i=1}^{d} \sigma_i(X_t^{x,\varepsilon})dW_t^i, \quad X_0^{x,\varepsilon} = x \in \mathbb{R}^N, \quad (4.47)$$

we have the following.

Corollary 4.5 (Small noise expansion) *For $m \geq 1$, there exists $C > 0$ such that*

$$\sup_{x\in\mathbb{R}^N} \left| \mathbb{E}[f(X_t^{x,\varepsilon})] - \left\{ \mathbb{E}[f(\bar{X}_t^{x,\varepsilon})] \right. \right.$$
$$+ \sum_{j=1}^{m} \varepsilon^j \sum_{k,\alpha^{(k)},\beta^{(k)},\gamma^{(p)}}^{(j)} \mathbb{E}\left[f(\bar{X}_t^{x,\varepsilon}) H_{\alpha^{(k)} * \gamma^{(p)}}(X_{1,t}^x, 1) \right]$$
$$\left. \frac{1}{p!} \left\langle DX_{1,t}^{x,\gamma_1} \otimes \cdots \otimes DX_{1,t}^{x,\gamma_p}, \mathbb{E}[D^p \prod_{e=1}^{k} X_{1+\beta_e,t}^{x,\alpha_e}] \right\rangle_{\mathcal{H}^{\otimes p}} \right\} \right|$$
$$\leq C \|f\|_\infty \varepsilon^{m+1}, \quad (4.48)$$

for any bounded measurable function $f : \mathbb{R}^N \to \mathbb{R}$ and $\varepsilon \in (0,1]$, where $X_{i,t}^x = \frac{1}{i!}\frac{\partial^i}{\partial \varepsilon^i}X_t^{x,\varepsilon}|_{\varepsilon=0}$, $i \in \mathbb{N}$ and $\bar{X}_t^{x,\varepsilon} = X_t^{x,0} + \varepsilon X_{1,t}^x$.

Proof of Corollary 4.5. Apply Corollary 4.2 and Theorem 4.2. □

Consider the elliptic diffusion:

$$dX_t^x = \sigma_0(X_t^x)dt + \sum_{i=1}^{d} \sigma_i(X_t^x)dW_t^i, \quad X_0^x = x \in \mathbb{R}^N, \quad (4.49)$$

we have the following two expansions.

4.5 Explicit Computation and Generalization

Corollary 4.6 (Small time expansion) *For $m \geq 1$, there exists $C > 0$ such that*

$$\sup_{x \in \mathbb{R}^N} \left| \mathbb{E}[f(X_t^x)] - \left\{ \mathbb{E}[f(\bar{X}_1^{x,\sqrt{t}})] \right. \right.$$

$$+ \sum_{j=1}^{m} t^{j/2} \sum_{k,\alpha^{(k)},\beta^{(k)},\gamma^{(p)}}^{(j)} \mathbb{E}[f(\bar{X}_1^{x,\sqrt{t}}) H_{\alpha^{(k)}}(\sum_{\ell=1}^{d} \sigma_\ell(x) W_1^\ell, H_{\gamma^{(p)}}(W_1, 1))]$$

$$\left. \left. \frac{1}{p!} \left\langle DW_1^{\gamma_1} \otimes \cdots \otimes DW_1^{\gamma_p}, \mathbb{E}[D^p \prod_{e=1}^{k} F_{\beta_e}^{x,\alpha_e}] \right\rangle_{\mathcal{H}^{\otimes p}} \right\} \right|$$

$$\leq C\|f\|_\infty t^{(m+1)/2}, \tag{4.50}$$

for any bounded measurable function $f : \mathbb{R}^N \to \mathbb{R}$ and $t \in (0, 1]$, where

$$F_0^x = \sum_{i=1}^{d} \sigma_i(x) W_1^i, \quad F_j^x = \sum_{\|\nu\|=j+1} \mathcal{L}_{\nu_1} \cdots \mathcal{L}_{\nu_{i-1}} \sigma_{\nu_i}(x) I_\nu(1), \quad i \in \mathbb{N}. \tag{4.51}$$

and $\bar{X}_1^{x,\varepsilon} = x + \varepsilon F_0^x = x + \varepsilon \sum_{i=1}^{d} \sigma_i(x) W_1^i$.

Proof of Corollary 4.6. Use $\mathbb{E}[f(\bar{X}_1^{x,\varepsilon}) G] = \int_{\mathbb{R}^d} f(x + \varepsilon \sigma(x) y) \mathbb{E}[\delta_y(W_1) G] dy$ and

$$\mathbb{E}[\delta_y(W_1) \prod_{e=1}^{k} F_{\beta_e}^{x,\alpha_e}]$$

$$= \sum_{\gamma^{(p)}, p \geq 0} \mathbb{E}[\delta_y(W_1) H_{\gamma^{(p)}}(W_1, 1)] \frac{1}{p!} \left\langle DW_1^{\gamma_1} \otimes \cdots \otimes DW_1^{\gamma_p}, \mathbb{E}[D^p \prod_{e=1}^{k} F_{\beta_e}^{x,\alpha_e}] \right\rangle_{\mathcal{H}^{\otimes p}}. \quad \square$$

Corollary 4.7 (Expansion around one-step Euler-Maruyama scheme) *For $m \geq 1$, there exists $C > 0$ such that*

$$\sup_{x \in \mathbb{R}^N} \left| \mathbb{E}[f(X_t^x)] - \left\{ \mathbb{E}[f(\bar{X}_t^{\text{EM},x})] \right. \right.$$

$$+ \sum_{j=1}^{m} \sum_{k,\alpha^{(k)},\beta^{(k)},\gamma^{(p)}}^{(j)} \mathbb{E}[f(\bar{X}_t^{\text{EM},x}) H_{\alpha^{(k)}}(\bar{X}_t^{\text{EM},x}, H_{\gamma^{(p)}}(W_t, 1))]$$

$$\left. \left. \frac{1}{p!} \left\langle DW_t^{\gamma_1} \otimes \cdots \otimes DW_t^{\gamma_p}, \mathbb{E}[D^p \prod_{e=1}^{k} X_{\beta_e,t}^{x,\alpha_e}] \right\rangle_{\mathcal{H}^{\otimes p}} \right\} \right|$$

$$\leq C\|f\|_\infty t^{(m+1)/2}$$

for any bounded measurable function $f : \mathbb{R}^N \to \mathbb{R}$ and $t \in (0, 1]$, where $\bar{X}_t^{\text{EM},x} = x + \sigma_0(x)t + \sum_{i=1}^d \sigma_i(x)W_t^i$ and

$$X_{j,t}^x = \sum_{|\nu|=j+1} \mathcal{L}_{\nu_1} \cdots \mathcal{L}_{\nu_j} \sigma_{\nu_{j+1}}(x) I_\nu(t), \quad j \in \mathbb{N}. \quad (4.52)$$

Proof of Corollary 4.7. Apply the similar calculation as in Corollary 4.6. □

Remark 4.6 We can extend the expansions in the section to rough differential equation driven by a d-dimensional fractional Brownian motion B^H with the Hurst index $H \in (1/3, 1/2)$. On a probability space $(\Omega, \mathcal{F}, \mathbb{P})$, let $\mathbf{B}^H$ be the canonical geometric rough path lift and consider the following rough differential equation:

$$dX_t^x = V_0(X_t^x)dt + V(X_t^x) \circ d\mathbf{B}_t^H \quad (4.53)$$

starting from $X_0^x = x \in \mathbb{R}^N$, where $V_0 \in C_b^\infty(\mathbb{R}^N; \mathbb{R}^N)$ and $V = (V_1, \ldots, V_d) \in C_b^\infty(\mathbb{R}^N; \mathbb{R}^{N \times d})$ (see Friz and Victoir 2009 or/and Friz and Hairer 2014 for more details on rough differential equations), and assume the uniformly elliptic condition. Then, one can show an explicit expansion of the probability distribution function as follows:

$$\mathbb{P}(X_t^x \leq y) = \mathbb{P}(\bar{X}_1^{x,t^H} \leq y)$$

$$+ t^H \mathbb{E}\left[\mathbf{1}_{\{\bar{X}_1^{x,t^H} \leq y\}} \sum_{j_1,j_2=1}^N \sum_{i_1,i_2,i_3=1}^d \hat{V}_{i_1} V_{i_2}^{j_1}(x) V_{i_3}^{j_2}(x) A_{j_1,j_2}(x) \right.$$

$$\frac{1}{2} \{ B_1^{H,i_1} B_1^{H,i_2} B_1^{H,i_3} - B_1^{H,i_1} \mathbf{1}_{i_2 = i_3 \neq 0} - B_1^{H,i_2} \mathbf{1}_{i_1 = i_3 \neq 0} \} \right]$$

$$+ t^{1-H} \mathbb{E}\left[\mathbf{1}_{\{\bar{X}_1^{x,t^H} \leq y\}} \sum_{j_1,j_2=1}^N \sum_{i_1=1}^d V_0^{j_1}(x) V_{i_1}^{j_2}(x) A_{j_1,j_2}(x) B_1^{H,i_1} \right] + O(t^{2H}) \quad (4.54)$$

where $\bar{X}_t^{x,\varepsilon} := x + \varepsilon F_t^0 = x + \varepsilon \sum_{i=1}^d V_i(x) B_t^{H,i}$, $x \in \mathbb{R}^N$, $t > 0$, $\varepsilon \in (0, 1]$, $\hat{V}_i f(x) = \sum_{j=1}^N V_i^j(x) \partial_j f(x), f \in C_b^\infty(\mathbb{R}^N), x \in \mathbb{R}^N$ and $A(x) = (A_{i,j}(x))_{1 \leq i,j \leq N}$ is the inverse matrix of $\sum_{i=1}^d V_i(x) \otimes V_i(x)$ for $x \in \mathbb{R}^N$. See Takahashi and Yamada (2024) for more details.

4.6 Notes and Summary

After the work of Watanabe (1987) and the studies of Kunitomo and Takahashi (1992), Kunitomo and Takahashi (2001), Kunitomo and Takahashi (2003), Takahashi (1995), Takahashi (1999), Yoshida (1992a), Yoshida (1992b) and Takahashi and Yoshida (2004), Takahashi and Yoshida 2005), asymptotic expansion schemes and their applications have been developed. See Takahashi (2015) and the references therein.

In this chapter, we provided a computational method of an asymptotic expansion and its numerical error:

$$|\mathbb{E}[f(F^\varepsilon)] - \{\mathbb{E}[f(F^0)] + \sum_{i=1}^{m} \varepsilon^i \mathbb{E}[f(F^0)\mathcal{H}_i]\}| = \|f\|_\infty \times O(\varepsilon^{m+1})$$

for a bounded measurable function f under a nondegenerate condition on F^0:

$$\|(\det \sigma^{F^0})^{-1}\|_p < \infty, \quad p \geq 1$$

as an extension of Takahashi and Yamada (2012), and particularly showed that how the general formula for Malliavin weights $\mathcal{H}_i$, $i \in \mathbb{N}$ is obtained.

We note that the method can be applied to various topics and targets such as stochastic volatility models, path-dependent Wiener functionals, backward stochastic differential equations and rough differential equations driven by fractional Brownian motions treated in Shiraya et al. (2012), Fujii and Takahashi (2019), Takahashi and Yamada (2015a), Takahashi and Yamada (2015b), Takahashi and Yamada (2024).

Chapter 5
Weak Approximation

Consider the solution of following SDE:

$$dX_t^x = \sigma_0(X_t^x)dt + \sum_{i=1}^{d} \sigma_i(X_t^x)dW_t^i, \quad X_0^x = x \in \mathbb{R}^N, \tag{5.1}$$

where $\sigma_i \in C_b^\infty(\mathbb{R}^N; \mathbb{R}^N)$, $i = 0, 1, \ldots, d$. Chapter 5 aims to obtain an efficient computation approach for the integral on the Wiener space:

$$\mathbb{E}[f(X_T^x)] = \int_{C_{(0)}([0,T];\mathbb{R}^d)} f(X_T^x(\omega))d\mu(\omega) \tag{5.2}$$

for an appropriate function $f : \mathbb{R}^N \to \mathbb{R}$ and $T > 0$ which is assumed to be not small.

We define a semigroup of linear operators $\{P_t\}_{t \geq 0}$ by

$$(P_t f)(x) = \mathbb{E}[f(X_t^x)], \quad t \geq 0, \; x \in \mathbb{R}^N, \; f \in \mathscr{B}_b(\mathbb{R}^N),$$

associated with the generator $\mathcal{L}$ (see Proposition 2.1).

If a family of operators $\{Q_t\}_{t \geq 0}$ given by

$$(Q_t f)(x) = \mathbb{E}[f(\hat{X}_t^x)], \quad t \geq 0, \; x \in \mathbb{R}^N, \; f \in \mathscr{B}_b(\mathbb{R}^N)$$

where $\{\hat{X}_t^x\}_{t \geq 0, x \in \mathbb{R}^N}$ be a stochastic process, satisfies

$$\|P_T f - (Q_{T/n})^n f\|_\infty = O\left(\frac{1}{n^m}\right), \tag{5.3}$$

we call the scheme $\{(Q_{T/n})^k\}_{1\leq k\leq n}$ (or the corresponding stochastic process) a m-order weak approximation of SDE.

5.1 From Euler-Maruyama to Higher-Order Weak Approximation

Let $\{Q_t^{EM}\}_{t\geq 0}$ be a family of operator

$$(Q_t^{EM} f)(x) = \mathbb{E}[f(\bar{X}_t^{EM,x})], \quad t \geq 0, \ x \in \mathbb{R}^N, \ f \in \mathscr{B}_b(\mathbb{R}^N),$$

with the one-step Euler-Maruyama scheme. Let $f \in C_b^\infty(\mathbb{R}^N)$. It is known that

$$\left\| P_t f - \left\{ f + \sum_{k=1}^m \frac{t^k}{k!} \mathcal{L}^k f \right\} \right\|_\infty = \|\nabla^{2(m+1)} f\|_\infty \times O(t^{m+1}) \tag{5.4}$$

(see Proposition 2.1 of Lyons and Victoir 2004) and

$$\|Q_t^{EM} f - \{f + t\mathcal{L} f\}\|_\infty = \|\nabla^4 f\|_\infty \times O(t^2) \tag{5.5}$$

(see Chap. 5 of Gobet 2020) and the proof of Theorem 7.7 of Pages (2018), for example). Thus, (with $m=1$ in (5.4)) it holds that

$$\|P_t f - Q_t^{EM} f\|_\infty = \|\nabla^4 f\|_\infty \times O(t^2), \tag{5.6}$$

and very roughly speaking, we have

$$\|P_T f - (Q_{T/n}^{EM})^n f\|_\infty = O\left(\frac{1}{n^2}\right) \times n = O\left(\frac{1}{n}\right),$$

which is proved even if f is not smooth whenever an elliptic condition is satisfied, using a technique of stochastic calculus. Here, $(Q_{T/n}^{EM})^n f$ is given by

$$(Q_{T/n}^{EM})^n f(x) = \mathbb{E}[f(\bar{X}_T^{EM,x,(n)})], \quad x \in \mathbb{R}^N,$$

where $\{\bar{X}_{kT/n}^{EM,x,(n)}\}_{k=0}^n$ is the (full) Euler-Maruyama scheme:

$$\bar{X}_{kT/n}^{EM,x,(n)} = \bar{X}_{(k-1)T/n}^{EM,x,(n)} + \sigma_0(\bar{X}_{(k-1)T/n}^{EM,x,(n)})T/n + \sum_{i=1}^d \sigma_i(\bar{X}_{(k-1)T/n}^{EM,x,(n)})\{W_{kT/n}^i - W_{(k-1)T/n}^i\},$$

$$\bar{X}_0^{EM,x,(n)} = x.$$

5.2 Third-Order Local Approximation

In principle, if a family of operators $\{Q_t\}_{t\geq 0}$ given by $(Q_t f)(x) = \mathbb{E}[f(\hat{X}_t^x)]$, $t \geq 0$, $x \in \mathbb{R}^N$, $f \in \mathscr{B}_b(\mathbb{R}^N)$ with a stochastic process $\{\hat{X}_t^x\}_{t\geq 0, x \in \mathbb{R}^N}$ satisfy

$$\left\| Q_t f - \left\{ f + \sum_{k=1}^{m} \frac{t^k}{k!} \mathcal{L}^k f \right\} \right\|_\infty = O(t^{m+1}), \tag{5.7}$$

for $f \in C_b^\infty(\mathbb{R}^N)$, one can construct a higher order weak approximation.

However, it is well known that such stochastic process and operators are difficult to implement when $d \geq 2$ and $m \geq 2$, due to the problem that iterated stochastic integrals of Brownian motion in stochastic Taylor expansion (i.e. $\omega \mapsto I_\alpha(t)(\omega)$, $\#\{i; \alpha \neq 0\} \geq 2$) such as Lévy areas are generally not continuous with respect to Brownian motion as Itô map and thus are not simulated. Whereas various schemes have been proposed in the literature (see the book of Kloeden and Platen 1992) for avoiding iterated stochastic integrals, the cost is to use additional random variables. One possible approach is Ninomiya-Victoir's scheme (Ninomiya and Victoir 2008) which provides an example for $m = 2$ using an algebraic scheme with the Kusuoka-Lyons-Victoir method (Kusuoka 2001 and Lyons and Victoir 2004), where simulation of Gaussian and Bernoulli random variables and a solver for an ordinary differential equation with the corresponding vector fields are required.

On the contrary, in Sect. 4.1, we have confirmed that the computation of asymptotic expansion automatically gives polynomialization of iterated stochastic integrals of Brownian motion through Skorohod integrals even when the target diffusion models are multidimensional. This fact implies that we can construct high-order local approximations using Brownian motion only, in other words, the computational cost for simulating random variables is the same as the Euler-Maruyama scheme. We show a new type of weak approximation of stochastic differential equations based on the asymptotic expansion method in Sect. 4.1.

While we do not take the strategy such as in the construction of Euler-Maruyama scheme above, (5.4), (5.5) and (5.6) suggest that one benefits from smoothness of a test function in the local error analysis in weak approximation. Thus, we make use of smoothness of the test function for the asymptotic expansion method and provide an original analysis for our construction of weak approximation.

5.2 Third-Order Local Approximation

We show a heuristic approach to constructing a local approximation with third-order small time error. Let us assume the uniformly elliptic condition. Then, we recall that the second-order expansion around one-step Euler-Maruyama scheme gives

$$\mathbb{E}[f(X_t^x)]$$
$$= \mathbb{E}[f(\bar{X}_t^{\text{EM},x})] + \mathbb{E}\left[f(\bar{X}_t^{\text{EM},x}) \sum_{i=1}^N H_{(i)}\left(\bar{X}_t^{\text{EM},x}, X_{1,t}^{x,i}\right)\right]$$
$$+ \mathbb{E}\left[f(\bar{X}_t^{\text{EM},x}) \sum_{i=1}^N H_{(i)}\left(\bar{X}_t^{\text{EM},x}, X_{2,t}^{x,i}\right)\right]$$
$$+ \mathbb{E}\left[f(\bar{X}_t^{\text{EM},x}) \sum_{i_1,i_2=1}^N \frac{1}{2} H_{(i_1,i_2)}\left(\bar{X}_t^{\text{EM},x}, \prod_{\ell=1}^2 X_{1,t}^{x,i_\ell}\right)\right] + \|f\|_\infty \times O(t^{3/2}),$$

where $X_{1,t}^{x,i} = \sum_{\alpha_1,\alpha_2=0}^d \mathcal{L}_{\alpha_1} \sigma_{\alpha_2}^i(x) I_{(\alpha_1,\alpha_2)}(t)$ and
$X_{2,t}^{x,i} = \sum_{\alpha_1,\alpha_2,\alpha_3=0}^d \mathcal{L}_{\alpha_1} \mathcal{L}_{\alpha_2} \sigma_{\alpha_3}^i(x) I_{(\alpha_1,\alpha_2,\alpha_3)}(t)$. Remarkably, although it gives the error *const.* $\|f\|_\infty \times t^{3/2}$, one can show that a smart approximation with a third order accuracy $O(t^3)$ in small time $t > 0$, by adding few terms to the first-order expansion or by reducing some terms of second-order expansion, if we can use smoothness of f. That is, we are easily able to obtain an approximation essentially using the first-order expansion, which provides the same accuracy as in Ninomiya-Victoir's scheme.

We first estimate the term:

$$\mathbb{E}\left[f(\bar{X}_t^{\text{EM},x}) \sum_{i=1}^N H_{(i)}\left(\bar{X}_t^{\text{EM},x}, X_{2,t}^{x,i}\right)\right]$$
$$= \sum_{i=1}^N \sum_{\alpha_1,\alpha_2,\alpha_3=0}^d \mathcal{L}_{\alpha_1} \mathcal{L}_{\alpha_2} \sigma_{\alpha_3}^i(x) \mathbb{E}\left[(\partial_i f)(\bar{X}_t^{\text{EM},x}) I_{(\alpha_1,\alpha_2,\alpha_3)}(t)\right].$$

Let $i = 1, \ldots, N$ and $\alpha \in \{0, 1, \ldots, d\}^3$.
If $\|\alpha\| = 3$,

$$\mathbb{E}\left[(\partial_i f)(\bar{X}_t^{\text{EM},x}) I_{(\alpha_1,\alpha_2,\alpha_3)}(t)\right]$$
$$= \sum_{\ell_1} \int_0^t \mathbb{E}\left[(\partial_i \partial_{\ell_1} f)(\bar{X}_t^{\text{EM},x}) \sigma_{\alpha_3}^{\ell_1}(x) I_{(\alpha_1,\alpha_2)}(t_1)\right] dt_1$$
$$= \sum_{\ell_1,\ell_2} \int_0^t \int_0^{t_1} \mathbb{E}\left[(\partial_i \partial_{\ell_1} \partial_{\ell_2} f)(\bar{X}_t^{\text{EM},x}) \sigma_{\alpha_3}^{\ell_1}(x) \sigma_{\alpha_2}^{\ell_2}(x) I_{(\alpha_1)}(t_2)\right] dt_2 dt_1$$
$$= \sum_{\ell_1,\ell_2,\ell_3} \int_0^t \int_0^{t_1} \int_0^{t_2} \mathbb{E}\left[(\partial_i \partial_{\ell_1} \partial_{\ell_2} \partial_{\ell_3} f), (\bar{X}_t^{\text{EM},x}) \sigma_{\alpha_3}^{\ell_1}(x) \sigma_{\alpha_2}^{\ell_2}(x) \sigma_{\alpha_2}^{\ell_3}(x)\right] dt_3 dt_2 dt_1$$
$$= \sum_{\ell_1,\ell_2,\ell_3} \mathbb{E}\left[(\partial_i \partial_{\ell_1} \partial_{\ell_2} \partial_{\ell_3} f)(\bar{X}_t^{\text{EM},x})\right] \frac{1}{6} t^3 \sigma_{\alpha_3}^{\ell_1}(x) \sigma_{\alpha_2}^{\ell_2}(x) \sigma_{\alpha_2}^{\ell_3}(x)$$

and then

5.2 Third-Order Local Approximation

$$\left|\mathbb{E}[(\partial_i f)(\bar{X}_t^{\text{EM},x})I_{(\alpha_1,\alpha_2,\alpha_3)}(t)]\right| = \|\nabla^4 f\|_\infty \times O(t^3).$$

If $\|\alpha\| = 4$,

$$\left|\mathbb{E}[(\partial_i f)(\bar{X}_t^{\text{EM},x})I_{(\alpha_1,\alpha_2,\alpha_3)}(t)]\right| = \|\nabla^3 f\|_\infty \times O(t^3).$$

If $\|\alpha\| = 5$,

$$\left|\mathbb{E}[(\partial_i f)(\bar{X}_t^{\text{EM},x})I_{(\alpha_1,\alpha_2,\alpha_3)}(t)]\right| = \|\nabla^2 f\|_\infty \times O(t^3).$$

If $\|\alpha\| = 6$,

$$\left|\mathbb{E}[(\partial_i f)(\bar{X}_t^{\text{EM},x})I_{(0,0,0)}(t)]\right| = \left|\mathbb{E}[(\partial_i f)(\bar{X}_t^{\text{EM},x})]\frac{1}{6}t^3\right| = \|\nabla f\|_\infty \times O(t^3).$$

These estimates show that if $|\alpha| = 3$,

$$\left|\mathbb{E}[(\partial_i f)(\bar{X}_t^{\text{EM},x})I_{(\alpha_1,\alpha_2,\alpha_3)}(t)]\right| = \|\nabla^{1+\#\{i;\alpha_i \neq 0\}} f\|_\infty \times O(t^3).$$

In general, we have the following estimate.

Lemma 5.1 *Let $k \in \mathbb{N}$ and $\alpha \in \{0, 1, \ldots, d\}^k$. There exists $C > 0$ such that for all $\varphi \in C_b^\infty(\mathbb{R}^N)$ and $t > 0$,*

$$\sup_{x \in \mathbb{R}^N} \left|\mathbb{E}[\varphi(\bar{X}_t^{\text{EM},x})I_\alpha(t)]\right| \leq C\|\nabla^{\#\{i;\alpha_i \neq 0\}}\varphi\|_\infty t^k.$$

Proof of Lemma 5.1. Since

$$\mathbb{E}[\varphi(\bar{X}_t^{\text{EM},x})I_\alpha(t)] = \int_{0<t_1<\cdots<t_k<t} \mathbb{E}[\partial^{\alpha_*}\varphi(\bar{X}_t^{\text{EM},x})]\prod_{k=1}^{|\alpha^*|}\sigma_{i_k}^{\alpha_k^*}(x)dt_1\cdots dt_k,$$

we have the assertion. □

Next we consider the term:

$$\mathbb{E}\left[f(\bar{X}_t^{\text{EM},x})\sum_{i_1,i_2=1}^{N}\frac{1}{2}H_{(i_1,i_2)}\left(\bar{X}_t^{\text{EM},x},\prod_{\ell=1}^{2}X_{t,1}^{x,i_\ell}\right)\right]$$

$$= \sum_{i_1,i_2=1}^{N}\frac{1}{2}\sum_{\alpha_1,\ldots,\alpha_4=0}^{d}\mathcal{L}_{\alpha_1}\sigma_{\alpha_2}^{i_1}(x)\mathcal{L}_{\alpha_3}\sigma_{\alpha_4}^{i_2}(x)\mathbb{E}[(\partial_{i_1}\partial_{i_2}f)(\bar{X}_t^{\text{EM},x})I_{(\alpha_1,\alpha_2)}(t)I_{(\alpha_3,\alpha_4)}(t)].$$

The Itô formula shows

$$I_{(\alpha_1,\alpha_2)}(t)I_{(\alpha_3,\alpha_4)}(t)$$
$$= I_{(\alpha_1,\alpha_2,\alpha_3,\alpha_4)}(t) + I_{(\alpha_1,\alpha_3,\alpha_2,\alpha_4)}(t) + I_{(\alpha_1,\alpha_3,\alpha_4,\alpha_2)}(t)$$
$$+ I_{(\alpha_3,\alpha_4,\alpha_1,\alpha_2)}(t) + I_{(\alpha_3,\alpha_1,\alpha_4,\alpha_2)}(t) + I_{(\alpha_3,\alpha_1,\alpha_2,\alpha_4)}(t)$$
$$+ I_{(\alpha_1,0,\alpha_4)}(t)\mathbf{1}_{\alpha_2=\alpha_3\neq 0} + I_{(\alpha_1,\alpha_3,0)}(t)\mathbf{1}_{\alpha_2=\alpha_4\neq 0} + I_{(0,\alpha_4,\alpha_2)}(t)\mathbf{1}_{\alpha_1=\alpha_3\neq 0}$$
$$+ I_{(0,\alpha_2,\alpha_4)}(t)\mathbf{1}_{\alpha_1=\alpha_3\neq 0} + I_{(\alpha_3,0,\alpha_2)}(t)\mathbf{1}_{\alpha_1=\alpha_4\neq 0} + I_{(\alpha_3,\alpha_1,0)}(t)\mathbf{1}_{\alpha_2=\alpha_4\neq 0}$$
$$+ I_{(0,0)}\mathbf{1}_{\alpha_1=\alpha_3\neq 0}\mathbf{1}_{\alpha_2=\alpha_4\neq 0}.$$

Note that if $|\alpha| = 3, 4$, there are $C > 0$ and $e \geq 1$ such that

$$\sup_{x\in\mathbb{R}^N} |\mathbb{E}[(\partial_{i_1}\partial_{i_2} f)(\bar{X}_t^{\mathrm{EM},x}) I_\alpha(t)]| \leq C \|\nabla^e f\|_\infty t^3$$

for all $t \in (0, 1]$. Therefore,

$$\mathbb{E}[(\partial_{i_1}\partial_{i_2} f)(\bar{X}_t^{\mathrm{EM},x}) I_{(\alpha_1,\alpha_2)}(t) I_{(\alpha_3,\alpha_4)}(t)]$$
$$= \mathbb{E}[(\partial_{i_1}\partial_{i_2} f)(\bar{X}_t^{\mathrm{EM},x}) I_{(0,0)}\mathbf{1}_{\alpha_1=\alpha_3\neq 0}\mathbf{1}_{\alpha_2=\alpha_4\neq 0}] + \|\nabla^e f\|_\infty \times O(t^3).$$

From the observation, we have

$$\mathbb{E}[f(X_t^x)]$$
$$= \mathbb{E}[f(\bar{X}_t^{\mathrm{EM},x})] + \mathbb{E}\left[f(\bar{X}_t^{\mathrm{EM},x}) \sum_{i=1}^N H_{(i)}\left(\bar{X}_t^{\mathrm{EM},x}, \sum_{\alpha_1,\alpha_2=0}^d \mathcal{L}_{\alpha_1}\sigma_{\alpha_2}^i(x) I_{(\alpha_1,\alpha_2)}(t)\right)\right]$$
$$+ \mathbb{E}\left[f(\bar{X}_t^{\mathrm{EM},x}) \sum_{i_1,i_2=1}^N \frac{1}{2} H_{(i_1,i_2)}\left(\bar{X}_t^{\mathrm{EM},x}, \sum_{\alpha_1,\alpha_2=1}^d \prod_{k=1}^2 \mathcal{L}_{\alpha_1}\sigma_{\alpha_2}^{i_k}(x) I_{(0,0)}(t)\right)\right]$$
$$+ \|\nabla^\ell f\|_\infty \times O(t^3)$$

for some $\ell \in \mathbb{N}$, which can be summarized and refined as follows.

Proposition 5.1 (Third-order local approximation) *For $f \in C_b^\infty(\mathbb{R}^N)$, we have*

$$\mathbb{E}[f(X_t^x)] = \mathbb{E}[f(\bar{X}_t^{\mathrm{EM},x})]$$
$$+ \sum_{i,j=1}^N \sum_{\alpha_1,\alpha_2=0}^d \sum_{\alpha_3=1}^d \frac{1}{2}\mathcal{L}_{\alpha_1}\sigma_{\alpha_2}^i(x)\frac{1}{t}A_{ij}(x)\sigma_{\alpha_3}^j(x) \mathbb{E}\left[f(\bar{X}_t^{\mathrm{EM},x})\mathbb{W}_{(\alpha_1,\alpha_2,\alpha_3)}(t)\right]$$
$$+ \sum_{i_1,i_2,j_1,j_2=1}^N \sum_{\alpha_1,\ldots,\alpha_4=1}^d \prod_{k=1}^2 \frac{1}{4}\mathcal{L}_{\alpha_1}\sigma_{\alpha_2}^{i_k}(x) A_{i_k j_k}(x)\sigma_{\alpha_2+k}^{j_k}(x) \mathbb{E}\left[f(\bar{X}_t^{\mathrm{EM},x})\mathbb{W}_{(\alpha_3,\alpha_4)}(t)\right]$$
$$+ R_{1,t}^f(x) + R_{2,t}^f(x)$$

with

5.2 Third-Order Local Approximation

$$\mathbb{W}_{(\alpha_1,\alpha_2,\alpha_3)}(t) = W_t^{\alpha_1} W_t^{\alpha_2} W_t^{\alpha_3} - W_t^{\alpha_1} t \mathbf{1}_{\alpha_2=\alpha_3\neq 0} - W_t^{\alpha_2} t \mathbf{1}_{\alpha_1=\alpha_3\neq 0} - W_t^{\alpha_3} t \mathbf{1}_{\alpha_1=\alpha_2\neq 0},$$
$$\mathbb{W}_{(\alpha_1,\alpha_2)}(t) = W_t^{\alpha_1} W_t^{\alpha_2} - t \mathbf{1}_{\alpha_1=\alpha_2\neq 0}, \quad \alpha_1,\alpha_2,\alpha_3 \in \{0,1,\ldots,d\},$$

where the map $(t,x) \mapsto R_{1,t}^f(x)$ satisfies that there is $C > 0$ independent of f such that $\sup_{x \in \mathbb{R}^N} |R_{1,t}^f(x)| \leq C \|f\|_\infty t^3$ for all $t \in (0,1]$, and $(t,x) \mapsto R_{2,t}^f(x)$ has the form $R_{2,t}^f(x) = t^3 \sum_{|\beta| \leq \ell} \mathbb{E}[(\partial^\beta f)(\bar{X}_t^{\mathrm{EM},x})] g_\beta(t,x)$ for some $\ell \in \mathbb{N}$ and bounded functions g_β, $\beta \in \{1,\ldots,N\}^k$, $k \leq \ell$, which is estimated by $\sup_{x \in \mathbb{R}^N} |R_{2,t}^f(x)| \leq \|\nabla^\ell f\|_\infty t^3$ for all $t \in (0,1]$.

Proof of Proposition 5.1. Corollary 4.4 with $m = 5$ gives

$$\mathbb{E}[f(X_t^x)] = \mathbb{E}[f(\bar{X}_t^{\mathrm{EM},x})]$$
$$+ \sum_{j=1}^{5} \sum_{k,\alpha^{(k)},\beta^{(k)}}^{(j)} \mathbb{E}[f(\bar{X}_t^{\mathrm{EM},x}) H_{\alpha^{(k)}}(\bar{X}_t^{\mathrm{EM},x}, \prod_{e=1}^{k} X_{\beta_e,t}^{x,\alpha_e})] + R_{1,t}^f(x),$$

and

$$\sup_{x \in \mathbb{R}^N} |R_{1,t}^f(x)| = \|f\|_\infty \times O(t^{(m+1)/2}).$$

Note that the expansion terms has the form:

$$\sum_{j=1}^{5} \sum_{k,\alpha^{(k)},\beta^{(k)}}^{(j)} \mathbb{E}[f(\bar{X}_t^{\mathrm{EM},x}) H_{\alpha^{(k)}}(\bar{X}_t^{\mathrm{EM},x}, \prod_{e=1}^{k} X_{\beta_e,t}^{x,\alpha_e})]$$
$$= \sum_{i=1}^{N} \mathbb{E}\left[\partial_i f(\bar{X}_t^{\mathrm{EM},x}) \sum_{\alpha_1,\alpha_2=0}^{d} \mathcal{L}_{\alpha_1} \sigma_{\alpha_2}^i(x) I_{(\alpha_1,\alpha_2)}(t)\right]$$
$$+ \sum_{i_1,i_2=1}^{N} \mathbb{E}\left[\partial_{i_1} \partial_{i_2} f(\bar{X}_t^{\mathrm{EM},x}) \frac{1}{2} \sum_{\alpha_1,\alpha_2=1}^{d} \prod_{k=1}^{2} \mathcal{L}_{\alpha_1} \sigma_{\alpha_2}^{i_k}(x) I_{(0,0)}(t)\right]$$
$$+ \sum_{\substack{2 \leq |\gamma| \leq 5, \\ 3 \leq |\beta| \leq 10}} \mathbb{E}[(\partial^\gamma f)(\bar{X}_t^{\mathrm{EM},x}) I_\beta(t)] c_{\beta,\gamma}(x)$$

for some $c_{\beta,\gamma} \in C_b^\infty(\mathbb{R}^N)$, $2 \leq |\gamma| \leq 5, 3 \leq |\beta| \leq 10$. Then, by Lemma 5.1, we have

$$\sum_{j=1}^{5}\sum_{k,\alpha^{(k)},\beta^{(k)}}^{(j)} \mathbb{E}[f(\bar{X}_t^{\text{EM},x})H_{\alpha^{(k)}}(\bar{X}_t^{\text{EM},x},\prod_{e=1}^{k}X_{\beta_e,t}^{x,\alpha_e})]$$

$$-\Big\{\mathbb{E}\Big[f(\bar{X}_t^{\text{EM},x})\sum_{i=1}^{N}H_{(i)}\Big(\bar{X}_t^{\text{EM},x},\sum_{\alpha_1,\alpha_2=0}^{d}\mathcal{L}_{\alpha_1}\sigma_{\alpha_2}^{i}(x)I_{(\alpha_1,\alpha_2)}(t)\Big)\Big]$$

$$+\mathbb{E}\Big[f(\bar{X}_t^{\text{EM},x})\sum_{i_1,i_2=1}^{N}\frac{1}{2}H_{(i_1,i_2)}\Big(\bar{X}_t^{\text{EM},x},\sum_{\alpha_1,\alpha_2=1}^{d}\prod_{k=1}^{2}\mathcal{L}_{\alpha_1}\sigma_{\alpha_2}^{i_k}(x)I_{(0,0)}(t)\Big)\Big]\Big\}$$

$$=t^3\sum_{|\beta|\leq\ell}\mathbb{E}[(\partial^\beta f)(\bar{X}_t^{\text{EM},x})]g_\beta(t,x).$$

We already checked that

$$\mathbb{E}\Big[f(\bar{X}_t^{\text{EM},x})\sum_{i=1}^{N}H_{(i)}\Big(\bar{X}_t^{\text{EM},x},\sum_{\alpha_1,\alpha_2=0}^{d}\mathcal{L}_{\alpha_1}\sigma_{\alpha_2}^{i}(x)I_{(\alpha_1,\alpha_2)}(t)\Big)\Big]$$

$$=\sum_{i,j=1}^{N}\sum_{\alpha_1,\alpha_2=0}^{d}\sum_{\alpha_3=1}^{d}\frac{1}{2}\mathcal{L}_{\alpha_1}\sigma_{\alpha_2}^{i}(x)\frac{1}{t}A_{ij}(x)\sigma_{\alpha_3}^{j}(x)\mathbb{E}\Big[f(\bar{X}_t^{\text{EM},x})\mathbb{W}_{(\alpha_1,\alpha_2,\alpha_3)}(t)\Big]$$

and an easy calculation gives

$$\mathbb{E}\Big[f(\bar{X}_t^{\text{EM},x})\sum_{i_1,i_2=1}^{N}\frac{1}{2}H_{(i_1,i_2)}\Big(\bar{X}_t^{\text{EM},x},\sum_{\alpha_1,\alpha_2=1}^{d}\prod_{k=1}^{2}\mathcal{L}_{\alpha_1}\sigma_{\alpha_2}^{i_k}(x)I_{(0,0)}(t)\Big)\Big]$$

$$=\sum_{i_1,i_2,j_1,j_2=1}^{N}\sum_{\alpha_1,\dots,\alpha_4=1}^{d}\prod_{k=1}^{2}\frac{1}{4}\mathcal{L}_{\alpha_1}\sigma_{\alpha_2}^{i_k}(x)A_{i_kj_k}(x)\sigma_{\alpha_{2+k}}^{j_k}(x)\mathbb{E}\Big[f(\bar{X}_t^{\text{EM},x})\mathbb{W}_{(\alpha_3,\alpha_4)}(t)\Big].\quad\Box$$

The result tells us that if the test function is smooth, one needs polynomials of Brownian motion up to the third-order in order to attain a third-order accuracy with respect to small time t, which leads to a second-order (global) weak approximation.

If $N=d$, the third-order local approximation is simply expressed by using the inverse matrix of σ as in the Bismut formula (Theorem 3.30).

Corollary 5.1 (Third-order local approximation (the case $N=d$)] *For $f\in C_b^\infty(\mathbb{R}^N)$, we have*

5.2 Third-Order Local Approximation

$$\mathbb{E}[f(X_t^x)] = \mathbb{E}[f(\bar{X}_t^{\mathrm{EM},x})]$$
$$+ \sum_{\alpha_1,\alpha_2=0}^{d} \sum_{\alpha_3,\alpha_4=1}^{d} \frac{1}{2t} \mathcal{L}_{\alpha_1} \sigma_{\alpha_2}^{\alpha_4}(x)(\sigma^{-1})_{\alpha_3\alpha_4}(x) \mathbb{E}\left[f(\bar{X}_t^{\mathrm{EM},x}) \mathbb{W}_{(\alpha_1,\alpha_2,\alpha_3)}(t)\right]$$
$$+ \sum_{\alpha_1,\ldots,\alpha_6=1}^{d} \frac{1}{4} \mathcal{L}_{\alpha_1} \sigma_{\alpha_2}^{\alpha_4}(x) \mathcal{L}_{\alpha_1} \sigma_{\alpha_2}^{\alpha_6}(x)(\sigma^{-1})_{\alpha_3\alpha_4}(x)(\sigma^{-1})_{\alpha_5\alpha_6} \mathbb{E}\left[f(\bar{X}_t^{\mathrm{EM},x}) \mathbb{W}_{(\alpha_3,\alpha_5)}(t)\right]$$
$$+ R_{1,t}^f(x) + R_{2,t}^f(x),$$

where $R_{1,t}^f$ and $R_{2,t}^f$ are the error functions satisfying $R_{i,t}^f(x) = O(t^3)$ for $x \in \mathbb{R}^d$, $i = 1, 2$ with the same properties on the bounds as in Proposition 5.1.

Proof of Corollary. In (4.36), we already checked through a divergence computation that

$$\sum_{i=1}^{N} \sum_{\alpha_1,\alpha_2=0}^{d} \mathcal{L}_{\alpha_1} \sigma_{\alpha_2}^{i}(x) \mathbb{E}\left[(\partial_i f)(\bar{X}_t^{\mathrm{EM},x}) I_{(\alpha_1,\alpha_2)}(t)\right]$$
$$= \sum_{i=1}^{N} \sum_{\alpha_1,\alpha_2=0}^{d} \mathcal{L}_{\alpha_1} \sigma_{\alpha_2}^{i}(x) \mathbb{E}[(\partial_i f)(\bar{X}_t^{\mathrm{EM},x}) \frac{1}{2} \mathbb{W}_{(\alpha_1,\alpha_2)}(t)].$$

Let $G = (G_1, \ldots, G_d)$ be $G_i = \mathcal{L}_{\alpha_1} \sigma_{\alpha_2}^{i}(x) \frac{1}{2}\{W_t^{\alpha_1} W_t^{\alpha_2} - t \mathbf{1}_{\{\alpha_1=\alpha_2\neq 0\}}\}$, $i = 1, \ldots, d$. Applying the chain rule of Malliavin derivative and the Malliavin integration by parts, we have

$$\mathbb{E}[\nabla f(\bar{X}_t^{\mathrm{EM},x}) G] = \frac{1}{t} E[\int_0^t \nabla f(\bar{X}_t^{\mathrm{EM},x}) \sigma(x) \sigma^{-1}(x) G ds]$$
$$= \frac{1}{t} \mathbb{E}[\int_0^t D_s f(\bar{X}_t^{\mathrm{EM},x}) \sigma^{-1}(x) G ds] = \frac{1}{t} \mathbb{E}[f(\bar{X}_t^{\mathrm{EM},x}) \sum_{k=1}^{d} \delta^k([\sigma^{-1}(x) G]^k \mathbf{1}_{\{\cdot \leq t\}})].$$

Then, the computation of the divergence operator gives

$$\sum_{k=1}^{d} \mathbb{E}[\partial_k f(\bar{X}_t^{\mathrm{EM},x}) G_k] = \sum_{k,\ell=1}^{d} \mathbb{E}\left[f(\bar{X}_t^{\mathrm{EM},x}) \frac{1}{t} (\sigma^{-1})_{k\ell}(x) \{G_\ell W_t^k - \int_0^t D_{k,s} G_\ell ds\}\right].$$

The same computation holds for the term:

$$\sum_{i_1,i_2=1}^{N} \frac{1}{2} \sum_{\alpha_1,\alpha_2=1}^{d} \prod_{\ell=1}^{2} \mathcal{L}_{\alpha_1} \sigma_{\alpha_2}^{i_\ell}(x) \mathbb{E}[(\partial_{i_1} \partial_{i_2} f)(\bar{X}_t^{\mathrm{EM},x}) I_{(0,0)}(t)].$$

Therefore the assertion holds. □

5.3 Second-Order Weak Approximation

Let $\{Q_t^{(2)}\}_{t>0}$ be a family of operators inspired by Proposition 5.1, given by

$$(Q_t^{(2)} f)(x) = \mathbb{E}[f(\bar{X}_t^{\text{EM},x})\mathcal{M}^{(2)}(t, x, W_t)],$$

for $t > 0$, $x \in \mathbb{R}^N$ and $f \in \mathscr{B}_b(\mathbb{R}^N)$, where

$$\mathcal{M}^{(2)}(t, x, W_t) = 1 + \sum_{i,j=1}^{N} \sum_{\alpha_1,\alpha_2=0}^{d} \sum_{\alpha_3=1}^{d} \frac{1}{2}\mathcal{L}_{\alpha_1}\sigma_{\alpha_2}^i(x)\frac{1}{t}A_{ij}(x)\sigma_{\alpha_3}^j(x)\mathbb{W}_{(\alpha_1,\alpha_2,\alpha_3)}(t)$$

$$+ \sum_{i_1,i_2,j_1,j_2=1}^{N} \sum_{\alpha_1,\ldots,\alpha_4=1}^{d} \frac{1}{4} \prod_{k=1}^{2} \mathcal{L}_{\alpha_1}\sigma_{\alpha_2}^{i_k}(x) A_{i_k j_k}(x) \sigma_{\alpha_{2+k}}^{j_k}(x) \mathbb{W}_{(\alpha_3,\alpha_4)}(t),$$

or if $N = d$,

$$\mathcal{M}^{(2)}(t, x, W_t) = 1 + \sum_{\alpha_1,\alpha_2=0}^{d} \sum_{\alpha_3,\alpha_4=1}^{d} \frac{1}{2t}\mathcal{L}_{\alpha_1}\sigma_{\alpha_2}^{\alpha_4}(x)(\sigma^{-1})_{\alpha_3\alpha_4}(x)\mathbb{W}_{(\alpha_1,\alpha_2,\alpha_3)}(t)$$

$$+ \sum_{\alpha_1,\ldots,\alpha_6=1}^{d} \frac{1}{4}\mathcal{L}_{\alpha_1}\sigma_{\alpha_2}^{\alpha_4}(x)\mathcal{L}_{\alpha_1}\sigma_{\alpha_2}^{\alpha_6}(x)(\sigma^{-1})_{\alpha_3\alpha_4}(x)(\sigma^{-1})_{\alpha_5\alpha_6}\mathbb{W}_{(\alpha_3,\alpha_5)}(t).$$

Then, we have the following.

Theorem 5.1 *It holds that*

$$\left\| P_T f - (Q_{T/n}^{(2)})^n f \right\|_\infty = O\left(\frac{1}{n^2}\right)$$

for any $f \in C_b^\infty(\mathbb{R}^N)$.

Proof of Theorem 5.1. By the definition of $\mathcal{M}^{(2)}(t, x, W_t)$, there exists $C > 0$ such that

$$\|Q_{T/n}^{(2)} f\|_\infty \le \|f\|_\infty \left(1 + \frac{C}{n}\right)$$

and

$$\|(Q_{T/n}^{(2)})^k f\|_\infty \le \|f\|_\infty e^C$$

5.3 Second-Order Weak Approximation

for all $n \geq 1$ and $k \leq n$. Then, we have

$$\|P_T f - (Q^{(2)}_{T/n})^n f\| \leq \sum_{k=0}^{n-1} \|(Q^{(2)}_{T/n})^k (P_{T/n} - Q^{(2)}_{T/n}) P_{T-(k+1)T/n} f\|_\infty$$

$$\leq C_1 \sum_{k=0}^{n-1} \|P_{T/n} P_{T-(k+1)T/n} f - Q^{(2)}_{T/n} P_{T-(k+1)T/n} f\|_\infty$$

$$\leq C_2 \sum_{k=0}^{n-1} \sum_{p=0}^{\ell} \|\nabla^p P_{T-(k+1)T/n} f\|_\infty \left(\frac{T}{n}\right)^3 \leq C_3 \sum_{k=0}^{n-1} \sum_{p=0}^{\ell} \|\nabla^p f\|_\infty \frac{1}{n^3} = C \sum_{p=0}^{\ell} \|\nabla^p f\|_\infty \frac{1}{n^2}$$

for some $C_1, C_2, C_3 > 0$ independent of $n \geq 1$. Here, we used the estimate

$$\sup_{x \in \mathbb{R}^N} \left| \frac{\partial^\alpha}{\partial x^\alpha} P_{T-t} f(x) \right| \leq C \sum_{e=1}^{k} \|\nabla^e f\|_\infty, \tag{5.8}$$

which is obtained by the computation:

$$\frac{\partial}{\partial x_i}(P_{T-t} f)(x) = \sum_{j=1}^{N} \mathbb{E}\left[(\partial_j f)(X^x_{T-t}) \frac{\partial}{\partial x_i} X^{x,j}_{T-t}\right]$$

and the iterative differentiations with the Kunita's estimate (Theorem 3.27). □

We now approximate $(P_T f)(x) = \mathbb{E}[f(X^x_T)]$ for a bounded measurable function $f : \mathbb{R}^N \to \mathbb{R}$. However, if f is bounded measurable only, we cannot use the smoothness as in (5.8). Actually, we have

$$\sup_{x \in \mathbb{R}^N} \left| \frac{\partial^\alpha}{\partial x^\alpha} P_{T-t} f(x) \right| \leq C \|f\|_\infty \frac{1}{(T-t)^{|\alpha|/2}}, \tag{5.9}$$

which is obtained by Kusuoka-Stroock's estimate ((3.29) in Theorem 3.31).

In the non-smooth case, the bound

$$\|\partial^\alpha P_{T-(k+1)T/n} f\|_\infty \leq C \|f\|_\infty \frac{1}{(T-(k+1)T/n)^{|\alpha|/2}} \tag{5.10}$$

will be huge as k becomes large (as n appears in the numerator of upper bound). Thus, we take another strategy to give the second order discretization with a bounded measurable test function. The trick is to use the Kusuoka-Stroock's estimate for the Eular-Maruyama scheme in a different way.

We have the following approximation.

Theorem 5.2 *There exists $C > 0$ such that*

$$\left\| P_T f - (Q^{(2)}_{T/n})^n f \right\|_\infty \leq C \|f\|_\infty \frac{1}{n^2}$$

for any $n \geq 1$ and bounded measurable function $f : \mathbb{R}^N \to \mathbb{R}$. In particular,

$$(Q_{T/n}^{(2)})^n f(x) = \mathbb{E}[f(\bar{X}_T^{\text{EM},x,(n)}) \prod_{i=1}^n \mathcal{M}^{(2)}(T/n, \bar{X}_{(i-1)T/n}^{\text{EM},x,(n)}, W_{iT/n} - W_{(i-1)T/n})],$$

$$x \in \mathbb{R}^N.$$

Proof of Theorem 5.2. The difference $P_T f - (Q_{T/n}^{(2)})^n f$ is decomposed as

$$P_T f - (Q_{T/n}^{(2)})^n f = \sum_{k=0}^{n-1} (Q_{T/n}^{(2)})^k (P_{T/n} - Q_{T/n}^{(2)}) P_{T-(k+1)T/n} f$$

$$= \sum_{k=0}^{n-1} (Q_{T/n}^{(2)})^k R_{1,T/n}^{P_{T-(k+1)T/n} f} + \sum_{k=0}^{n-1} (Q_{T/n}^{(2)})^k R_{2,T/n}^{P_{T-(k+1)T/n} f}.$$

We immediately have

$$\Big\| \sum_{k=0}^{n-1} (Q_{T/n}^{(2)})^k R_{1,T/n}^{P_{T-(k+1)T/n} f} \Big\|_\infty \leq \sum_{k=0}^{n-1} C \| P_{T-(k+1)T/n} f \|_\infty \frac{1}{n^3}$$

$$\leq C \| f \|_\infty \sum_{k=0}^{n-1} \frac{1}{n^3} = C \| f \|_\infty \frac{1}{n^2}.$$

For the function $\sum_{k=0}^{n-1} (Q_{T/n}^{(2)})^k R_{2,T/n}^{P_{T-(k+1)T/n} f}$, we decompose it as

$$\sum_{k=0}^{n-1} (Q_{T/n}^{(2)})^k R_{2,T/n}^{P_{T-(k+1)T/n} f}(x)$$

$$= \sum_{(k+1)T/n < [T/2]} (Q_{T/n}^{(2)})^k R_{2,T/n}^{P_{T-(k+1)T/n} f}(x) + \sum_{(k+1)T/n \geq [T/2]} (Q_{T/n}^{(2)})^k R_{2,T/n}^{P_{T-(k+1)T/n} f}(x)$$

$$= \sum_{(k+1)T/n < [T/2]} \sum_{|\alpha| \leq \ell} \mathbb{E}[(\partial^\alpha P_{T-(k+1)T/n} f)(\bar{X}_{(k+1)T/n}^{\text{EM},x,(n)}) G_{\alpha,k}^{x,(n)}] \frac{T^3}{n^3}$$

$$+ \sum_{(k+1)T/n \geq [T/2]} \sum_{|\alpha| \leq \ell} \mathbb{E}[P_{T-(k+1)T/n} f(\bar{X}_{(k+1)T/n}^{\text{EM},x,(n)}) H_\alpha(\bar{X}_{(k+1)T/n}^{\text{EM},x,(n)}, G_{\alpha,k}^{x,(n)})] \frac{T^3}{n^3},$$

where $G_{\alpha,k}^{x,(n)} \in \mathbb{D}^\infty$, $|\alpha| \leq \ell$, $n \geq 1$, $k \leq n$ given by

$$G_{\alpha,k}^{x,(n)} = [\prod_{i=1}^k \mathcal{M}^{(2)}(T/n, \bar{X}_{(i-1)T/n}^{\text{EM},x,(n)}, W_{iT/n} - W_{(i-1)T/n})] g_\alpha(T/n, \bar{X}_{kT/n}^{\text{EM},x,(n)}),$$

5.3 Second-Order Weak Approximation

for $|\alpha| \leq \ell, n \geq 1, k \leq n$, satisfy that for all $r \geq 1$, $p \in [1, \infty)$, there is $C > 0$ such that $\sup_{x \in \mathbb{R}^N} \|G_{\alpha,k}^{x,(n)}\|_{r,p} \leq C$ for all $|\alpha| \leq \ell, n \geq 1, k \leq n$. We apply the Kusuoka-Stroock estimate (3.29) in Theorem 3.31 to provide the desired sharp upper bound, and obtain the following lemma.

Lemma 5.2 *Let α be a multi-index.*

1. *There exists $C > 0$ such that*

$$\sup_{(k+1)T/n < [T/2]} \sup_{x \in \mathbb{R}^N} \left|\mathbb{E}[(\partial^\alpha P_{T-(k+1)T/n} f)(\bar{X}_{(k+1)T/n}^{\mathrm{EM},x,(n)}) G_{\alpha,k}^{x,(n)}]\right| \leq C\|f\|_\infty \frac{1}{T^{|\alpha|/2}} \tag{5.11}$$

for all $n \geq 1$.

2. *There exists $C > 0$ such that*

$$\sup_{(k+1)T/n \geq [T/2]} \sup_{x \in \mathbb{R}^N} \left|\mathbb{E}[P_{T-(k+1)T/n} f(\bar{X}_{(k+1)T/n}^{\mathrm{EM},x,(n)}) H_\alpha(\bar{X}_{(k+1)T/n}^{\mathrm{EM},x,(n)}, G_{\alpha,k}^{x,(n)})]\right|$$
$$\leq C\|f\|_\infty \frac{1}{T^{|\alpha|/2}} \tag{5.12}$$

for all $n \geq 1$.

Proof of Lemma 5.2.

1. Recalling that by the Kusuoka-Stroock estimate for the solution of SDE X^x,

$$\|\partial^\alpha P_{T-(k+1)T/n} f\|_\infty$$
$$= \sup_{x \in \mathbb{R}^N} \left|\frac{\partial^\alpha}{\partial x^\alpha} \mathbb{E}[f(X_{T-(k+1)T/n}^x)]\right| \leq C_1 \|f\|_\infty \frac{1}{(T-(k+1)T/n)^{|\alpha|/2}}, \tag{5.13}$$

we have

$$\sup_{(k+1)T/n < [T/2]} \sup_{x \in \mathbb{R}^N} \left|\mathbb{E}[(\partial^\alpha P_{T-(k+1)T/n} f)(\bar{X}_{(k+1)T/n}^{\mathrm{EM},x,(n)}) G_{\alpha,k}^{x,(n)}]\right| \leq C_2 \|f\|_\infty \frac{1}{T^{|\alpha|/2}} \tag{5.14}$$

for some $C_1, C_2 > 0$ independent of $n \geq 1$.

2. Using the Kusuoka-Stroock estimate for the Euler-Maruyama scheme $\bar{X}^{\mathrm{EM},x,(n)}$ as an elliptic Itô process:

$$\bar{X}_t^{\mathrm{EM},x,(n)} = x + \int_0^t \sigma_0(\bar{X}_{\psi(s)}^{\mathrm{EM},x,(n)}) ds + \sum_{i=1}^d \int_0^t \sigma_i(\bar{X}_{\psi(s)}^{\mathrm{EM},x,(n)}) dW_s^i, \quad t \geq 0, \tag{5.15}$$

where $\psi(s) = \sup\{kT/n;\ kT/n \leq s\}$, $s \in [0, T]$, there exists $C > 0$ such that

$$\left|\mathbb{E}[P_{T-(k+1)T/n}f(\bar{X}^{\mathrm{EM},x,(n)}_{(k+1)T/n})H_\alpha(\bar{X}^{\mathrm{EM},x,(n)}_{(k+1)T/n}, G^{x,(n)}_{\alpha,k})]\right|$$
$$\leq C\|P_{T-(k+1)T/n}f\|_\infty \frac{1}{T^{|\alpha|/2}} \leq C\|f\|_\infty \frac{1}{T^{|\alpha|/2}} \tag{5.16}$$

for all $x \in \mathbb{R}^N$, $n \geq 1$ and $k \leq n$ such that $(k+1)T/n \geq [T/2]$. □

Then, by Lemma 5.2, we have

$$\left\|\sum_{k=0}^{n-1}(Q^{(2)}_{T/n})^k R^{P_{T-(k+1)T/n}f}_{2,T/n}\right\|_\infty \leq C\|f\|_\infty \sum_{k=0}^{n-1}\frac{1}{n^3} = C\|f\|_\infty \frac{1}{n^2}.$$

Therefore, we finally get the estimate

$$\|P_T f - (Q^{(2)}_{T/n})^n f\|_\infty \leq C\|f\|_\infty \frac{1}{n^2},$$

and by construction of $\{Q^{(2)}_t\}_{t>0}$, we have the representation:

$$(Q^{(2)}_{T/n})^n f(x) = \mathbb{E}[f(\bar{X}^{\mathrm{EM},x,(n)}_T)\prod_{k=1}^n \mathcal{M}^{(2)}(T/n, \bar{X}^{\mathrm{EM},x,(n)}_{(k-1)T/n}, W_{kT/n} - W_{(k-1)T/n})],$$

$$x \in \mathbb{R}^N.$$

□

5.4 General High Order Weak Approximation

The third order local approximation (Proposition 5.1) can be generalized as follows.

Proposition 5.2 *Let $f \in C_b^\infty(\mathbb{R}^N; \mathbb{R})$. We have*

$$\mathbb{E}[f(X^x_t)] = \mathbb{E}[f(\bar{X}^{\mathrm{EM},x}_t)] + \sum_{j=1}^m \sum_{\substack{k,\alpha^{(k)},\beta^{(k)},\gamma^{(p)} \\ j+k+p \leq 2m}}^{(j)} \mathbb{E}[f(\bar{X}^{\mathrm{EM},x}_t)H_{\alpha(k)}(\bar{X}^{\mathrm{EM},x}_t, H_{\gamma(p)}(W_t, 1))]$$

$$\frac{1}{p!}\left\langle DW^{\gamma_1}_t \otimes \cdots \otimes DW^{\gamma_p}_t, \mathbb{E}[D^p \prod_{e=1}^k X^{x,\alpha_e}_{\beta_e,t}]\right\rangle_{\mathcal{H}^{\otimes p}}$$

$$+ R^{m+1,f}_{1,t}(x) + R^{m+1,f}_{2,t}(x),$$

5.4 General High Order Weak Approximation

where

$$\sum_{\substack{k,\alpha^{(k)},\beta^{(k)},\gamma^{(p)} \\ j+k+p\leq 2m}}^{(j)} = \sum_{\substack{\beta^{(k)}\in \mathbb{N}^k, \sum_{\ell=1}^k \beta_\ell = j, \\ \alpha^{(k)} \in \{1,\ldots,N\}^k, 1\leq k\leq j, \gamma^{(p)}\in\{1,\ldots,N\}^p, p\geq 0 \\ j+k+p\leq 2m}} \frac{1}{k!},$$

$$X_{j,t}^x = \sum_{|\nu|=j+1} \mathcal{L}_{\nu_1}\cdots\mathcal{L}_{\nu_j}\sigma_{\nu_{j+1}}(x) I_\nu(t), \quad j \in \mathbb{N},$$

and $R_{i,\cdot}^{m+1,f}$, $i = 1, 2$ satisfy that there exist $C_1, C_2 > 0$ and $\ell(m) \in \mathbb{N}$ which are independent of f such that

$$\sup_{x\in\mathbb{R}^N} |R_{1,t}^{m+1,f}(x)| \leq C\|f\|_\infty t^{m+1} \quad \text{and} \quad \sup_{x\in\mathbb{R}^N} |R_{2,t}^{m+1,f}(x)| \leq C\|\nabla^{\ell(m)} f\|_\infty t^{m+1}$$

for all $t > 0$.

Proof of Proposition 5.2. Recalling Corollary 4.7, for $j \in \mathbb{N}$, we have

$$\sum_{k,\alpha^{(k)},\beta^{(k)},\gamma^{(p)}}^{(j)} \mathbb{E}[f(\bar{X}_t^{\text{EM},x}) H_{\alpha^{(k)}}(\bar{X}_t^{\text{EM},x}, H_{\gamma^{(p)}}(W_t, 1))]$$

$$\times \frac{1}{p!}\Big\langle DW_t^{\gamma_1} \otimes \cdots \otimes DW_t^{\gamma_p}, \mathbb{E}[D^p \prod_{e=1}^k X_{\beta_e,t}^{x,\alpha_e}]\Big\rangle_{\mathcal{H}^{\otimes p}}$$

$$= \sum_{k,\alpha^{(k)},\beta^{(k)}}^{(j)} \mathbb{E}\Big[f(\bar{X}_t^{\text{EM},x}) H_{\alpha^{(k)}}(\bar{X}_t^{\text{EM},x}, \prod_{e=1}^k X_{\beta_e,t}^{x,\alpha_e})\Big]$$

$$= \sum_{k,\alpha^{(k)},\beta^{(k)}}^{(j)} \mathbb{E}\Big[\partial^{\alpha^{(k)}} f(\bar{X}_t^{\text{EM},x}) \prod_{e=1}^k X_{\beta_e,t}^{x,\alpha_e}\Big]$$

$$= \sum_{k,\alpha^{(k)},\beta^{(k)},\gamma^{(p)}}^{(j)} \mathbb{E}\Big[\partial^{\alpha^{(k)} * \gamma^{(p)}} f(\bar{X}_t^{\text{EM},x})\Big]$$

$$\times \frac{1}{p!}\Big\langle D\bar{X}_t^{\text{EM},x,\gamma_1} \otimes \cdots \otimes D\bar{X}_t^{\text{EM},x,\gamma_p}, \mathbb{E}[D^p \prod_{e=1}^k X_{\beta_e,t}^{x,\alpha_e}]\Big\rangle_{\mathcal{H}^{\otimes p}}.$$

For $j = 1,\ldots,m$, $k = 1,\ldots,j$, $\beta_1,\ldots,\beta_k \geq 1$ such that $\sum_{\ell=1}^k \beta_\ell = j$, $(\alpha_1,\ldots,\alpha_k) \in \{1,\ldots,N\}^k$, $(\gamma_1,\ldots,\gamma_p) \in \{1,\ldots,N\}^p$, $p \geq 0$ such that $j + k + p \geq 2m + 1$, there exists $C > 0$ such that

$$\sup_{x\in\mathbb{R}^N} \Big|\mathbb{E}\Big[\partial^{\alpha^{(k)} * \gamma^{(p)}} f(\bar{X}_t^{\text{EM},x})\Big] \frac{1}{p!}\Big\langle D\bar{X}_t^{\text{EM},x,\gamma_1} \otimes \cdots \otimes D\bar{X}_t^{\text{EM},x,\gamma_p}, \mathbb{E}[D^p \prod_{e=1}^k X_{\beta_e,t}^{x,\alpha_e}]\Big\rangle_{\mathcal{H}^{\otimes p}}\Big|$$

$$\leq C\|\nabla^{k+p} f\|_\infty t^{m+1}. \tag{5.17}$$

□

Let

$$\mathcal{M}^{(m)}(t,x,W_t) = 1 + \sum_{\substack{j=1 \\ j+k+p \leq 2m}}^{m} \sum_{k,\alpha^{(k)},\beta^{(k)},\gamma^{(p)}}^{(j)} H_{\alpha^{(k)}}(\sigma(x)W_t, H_{\gamma^{(p)}}(W_t, 1))$$

$$\times \frac{1}{p!} \Big\langle DW_t^{\gamma_1} \otimes \cdots \otimes DW_t^{\gamma_p}, \mathbb{E}[D^p \prod_{e=1}^{k} X_{\beta_e,t}^{x,\alpha_e}] \Big\rangle_{\mathcal{H}^{\otimes p}}$$

(5.18)

and

$$(Q_t^{(m)} f)(x) := \mathbb{E}[f(\bar{X}_t^{\mathrm{EM},x}) \mathcal{M}^{(m)}(t,x,W_t)], \quad x \in \mathbb{R}^N, \ t > 0, \ f \in \mathcal{B}_b(\mathbb{R}^N).$$

We have the following high order weak approximation.

Theorem 5.3 *For $m \in \mathbb{N}$, there exists $C > 0$ such that*

$$\|P_T f - (Q_{T/n}^{(m)})^n f\|_\infty \leq C \|f\|_\infty \frac{1}{n^m}$$

for any bounded measurable function $f : \mathbb{R}^N \to \mathbb{R}$ and $n \geq 1$. In particular,

$$(Q_{T/n}^{(m)})^n f(x) = \mathbb{E}[f(\bar{X}_T^{\mathrm{EM},x,(n)}) \prod_{i=1}^{n} \mathcal{M}^{(m)}(T/n, \bar{X}_{(i-1)T/n}^{\mathrm{EM},x,(n)}, W_{iT/n} - W_{(i-1)T/n})],$$

$$x \in \mathbb{R}^N.$$

Proof of Theorem 5.3. Apply the similar proof of Theorem 5.2. □

5.5 Numerical Recipe

We introduce the algorithm of the weak approximation scheme implemented with Monte-Carlo (MC) and Quasi Monte-Carlo (QMC) methods by comparing it with that of the Euler-Maruyama scheme. Below, $\mathcal{N}(0,1)$ and Φ represent the standard Gaussian random variable and its distribution function, and let $t_i = iT/n$, $i = 0, 1, \ldots, n$ for $n \geq 1$.

5.5 Numerical Recipe

Algorithm 5.1 Euler-Maruyama scheme

for $j = 1$ to M do
  $\bar{X}_{t_0}^{\text{EM},(n),[j]} = x$
  for $i = 1$ to n do
    Simulate i.i.d. $Z^k \sim \mathcal{N}(0, 1), k = 1, \ldots, d$
    Update
    $$\bar{X}_{t_i}^{\text{EM},(n),[j]} = \bar{X}_{t_{i-1}}^{\text{EM},(n),[j]} + \sigma_0(\bar{X}_{t_{i-1}}^{\text{EM},(n),[j]})T/n + \sum_{k=1}^{d}\sigma_k(\bar{X}_{t_{i-1}}^{\text{EM},(n),[j]})\sqrt{T/n}Z^k$$
  end for
end for
Return $\frac{1}{M}\sum_{j=1}^{M} f(\bar{X}_T^{\text{EM},(n),[j]})$

Algorithm 5.2 m-order weak approximation scheme (MC)

for $j = 1$ to M do
  $\bar{X}_{t_0}^{\text{EM},(n),[j]} = x, \mathcal{W}_{t_0}^{[j]} = 1$
  for $i = 1$ to n do
    Simulate $Z = (Z^1, \ldots, Z^k)$ with i.i.d. $Z^k \sim \mathcal{N}(0, 1), k = 1, \ldots, d$
    Update Malliavin weight $\mathcal{W}_{t_i}^{[j]} = \mathcal{W}_{t_{i-1}}^{[j]} \times \mathcal{M}^{(m)}(T/n, \bar{X}_{t_{i-1}}^{\text{EM},(n),[j]}, \sqrt{T/n}Z)$
    Update
    $$\bar{X}_{t_i}^{\text{EM},(n),[j]} = \bar{X}_{t_{i-1}}^{\text{EM},(n),[j]} + \sigma_0(\bar{X}_{t_{i-1}}^{\text{EM},(n),[j]})T/n + \sum_{k=1}^{d}\sigma_k(\bar{X}_{t_{i-1}}^{\text{EM},(n),[j]})\sqrt{T/n}Z^k$$
  end for
end for
Return $\frac{1}{M}\sum_{j=1}^{M} f(\bar{X}_T^{\text{EM},(n),[j]})\mathcal{W}_T^{[j]}$

Algorithm 5.3 m-order weak approximation scheme (QMC)

Generate $\xi_i^j = (\xi_{i,1}^j, \ldots, \xi_{i,d}^j)$, $\xi_{i,k}^j = \Phi^{-1}(y_{(i-1)d+k}^j)$, $1 \leq i \leq n$, $1 \leq k \leq d$ from $(n \times d)$-dimensional vector $y^j = (y_1^j, \ldots, y_{n \times d}^j)$ from a low discrepancy sequence, $j = 1, \ldots, M$
for $j = 1$ to M do
  $\bar{X}_{t_0}^{\text{EM},(n),[j]} = x, \mathcal{W}_{t_0}^{[j]} = 1$
  for $i = 1$ to n do
    Update the weight $\mathcal{W}_{t_i}^{[j]} = \mathcal{W}_{t_{i-1}}^{[j]} \times \mathcal{M}^{(m)}(T/n, \bar{X}_{t_{i-1}}^{\text{EM},(n),[j]}, \sqrt{T/n}\xi_i^j)$
    Update
    $$\bar{X}_{t_i}^{\text{EM},(n),[j]} = \bar{X}_{t_{i-1}}^{\text{EM},(n),[j]} + \sigma_0(\bar{X}_{t_{i-1}}^{\text{EM},(n),[j]})T/n + \sum_{k=1}^{d}\sigma_k(\bar{X}_{t_{i-1}}^{\text{EM},(n),[j]})\sqrt{T/n}\xi_{i,k}^j$$
  end for
end for
Return $\frac{1}{M}\sum_{j=1}^{M} f(\bar{X}_T^{\text{EM},(n),[j]})\mathcal{W}_T^{[j]}$

As an example, we show a Python code for computing the integral:

$$\mathbb{E}[\max\{X_T^1 - K, 0\}] = \int_{C_{(0)}([0,T];\mathbb{R}^2)} \max\{X_T^1(\omega) - K, 0\} d\mu(\omega) \quad (5.19)$$

with the first and second order weak approximation scheme under the lognormal SABR model:

$$dX_t^1 = X_t^1 X_t^2 dW_t^1, \; X_0^1 = x_1,$$
$$dX_t^2 = \nu X_t^2 [\rho dW_t^1 + \sqrt{1-\rho^2} dW_t^2], \; X_0^1 = x_2. \quad (5.20)$$

Listing 5.1 SABR_WA_QMC.py

```
from scipy.stats import qmc
from scipy.stats import norm
import numpy as np
import time

def g(x,K):
    return np.maximum(x - K, 0.0)

def SABR_WA_QMC(M, n, x1, x2, K, T, r, nu, rho, WA_order):

    def M_weight(x1, x2, w0, w1, w2, WA_order):

        if WA_order==1:
            return 1.0
        elif WA_order==2:
            x2 = np.log(x2)
            inv11 = 1.0 / (np.exp(x2))
            inv21 = -nu * rho / (np.exp(x2)*nu*np.sqrt(1.0 -
                rho**2))

            L0V01 = r**2
            L0V11 = r*np.exp(x2) + np.exp(2.0 * x2)*rho*nu
            L1V01 = r*np.exp(x2)
            L1V11 = np.exp(2.0 * x2) + np.exp(x2)*rho*nu
            L2V11 = np.exp(x2)*np.sqrt(1.0 - rho**2)*nu

            w00 = w0 * w0
            w01 = w0 * w1
            w02 = w0 * w2
            w11 = w1 * w1 - w0
            w12 = w1 * w2
            w22 = w2 * w2 - w0
            w111 = w1**3 - 3.0 * w1*w0
            w001 = w1 * w00
            w002 = w2 * w00
            w011 = w11 * w0
            w112 = w11 * w2
            w122 = w22 * w1
            w012 = w0*w1*w2

            A = 1.0/(2.0*w0)*(
                inv11 * (L0V01 * w001 + L0V11 * w011 + L1V01 *
                    w011 + L1V11 * w111 + L2V11 * w112)
                +inv21 * (L0V01 * w002 + L0V11 * w012 + L1V01 *
                    w012 + L1V11 * w112 + L2V11 * w122)
            )
```

5.5 Numerical Recipe

```
44
45                  B = 1.0/4.0*(
46                      inv11 * inv11 * (L1V11 * L1V11* w11 + L2V11 *
                            L2V11* w11)
47                      + inv11 * inv21 * (L1V11 * L1V11* w12 + L2V11 *
                            L2V11* w12) * 2.0
48                      + inv21 * inv21 * (L1V11 * L1V11* w22 + L2V11 *
                            L2V11* w22)
49                  )
50
51                  return 1.0 + A + B
52              else:
53                  return None
54
55          Weight = 1.0
56          X1 = np.ones ((M, 1)) * x1
57          X2 = np.ones ((M, 1)) * x2
58          qrng = qmc.Sobol(d=2*n, scramble=True)
59          sample = qrng.random(M)
60
61          for _n in range (n):
62              dW1 = np.reshape(norm.ppf(sample[:,(2*_n):(2*_n+1)]), (
                      M, 1))*np.sqrt(T/n)
63              dW2 = np.reshape(norm.ppf(sample[:,(2*_n+1):(2*_n+2)]),
                      (M, 1))*np.sqrt(T/n)
64              Weight = Weight * M_weight(X1, X2, T/n, dW1, dW2,
                      WA_order)
65
66              X1 = X1 + r * X1 * T/n + X2*X1*dW1
67              X2 = X2 * np.exp(-1.0/2.0*nu**2*T/n + nu*rho*dW1 + nu*
                      np.sqrt(1.0-rho**2)*dW2)
68              X1 = np.maximum(X1,0.0)
69          return np.exp(-r*T)*np.mean(g(X1, K)*Weight)
70
71   M = 10**7
72   x1, x2, K, T, r, nu, rho = 100.0, 0.2, 100.0, 1.0, 0.0, 0.1,
         -0.5
73
74   for order in [1,2]:
75       for n in [1,2,4,8,16,32]:
76           t0 = time.time()
77           value = SABR_WA_QMC(M, n, x1, x2, K, T, r, nu, rho,
                 order)
78           t1 = time.time()
79           print("WA of order
```

The accuracy of the weak approximation can be checked through a numerical example for SABR stochastic volatility model using the algorithm (Fig. 5.1).

Remark 5.1 We compare the proposed scheme with the *Ninomiya-Victoir* (NV) scheme introduced by Ninomiya and Victoir (2008), which is well-known for simple implementation of a second order weak method.

The NV scheme is based on the following one-step approximation:

$$\bar{X}_t^{\text{NV}}(x) = \begin{cases} \exp(\frac{t}{2}V_0) \circ \exp(\sqrt{t}Z^1 V_1) \circ \cdots \circ \exp(\sqrt{t}Z^d V_d) \circ \exp(\frac{t}{2}V_0)(x), & \text{if } N = 1, \\ \exp(\frac{t}{2}V_0) \circ \exp(\sqrt{t}Z^d V_d) \circ \cdots \circ \exp(\sqrt{t}Z^1 V_1) \circ \exp(\frac{t}{2}V_0)(x), & \text{if } N = -1, \end{cases} \quad (5.21)$$

where N is a Bernoulli random variable with the distribution $P(N = 1) = P(N = -1) = 1/2$ and $\{Z^i\}_{i=1,\ldots,d}$ is a family of i.i.d $Z^i \sim \mathcal{N}(0, 1), i = 1, \ldots, d$ such that

Fig. 5.1 Weak approximation error

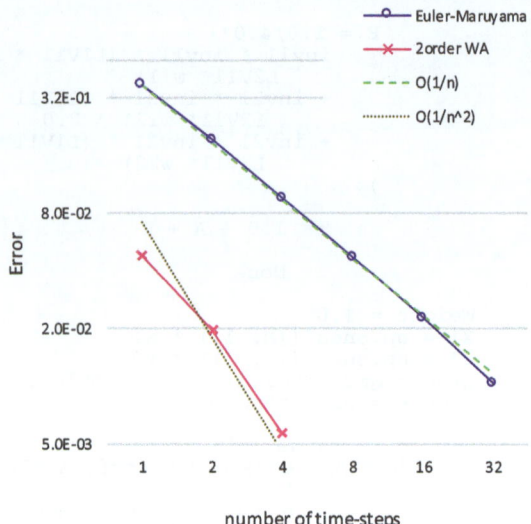

N and $\{Z^i\}_{i=1,\ldots,d}$ are independent. Then NV scheme is given by $(Q_{T/n}^{NV})^n f$, where $(Q_t^{NV})_t$ is a local operator defined by $(Q_t^{NV}\varphi)(x) = E[\varphi(\bar{X}_t^{NV}(x))]$ through $(d+1)$-dimensional random variable $(N, Z^1, \ldots, Z^d)$. The features of the proposed scheme is different from that of NV scheme in the following sense:

- NV scheme need to solve an $(n \times (d+1))$-dimensional integral by QMC whose dimension is higher than ours since the proposed scheme always uses an $(n \times d)$-dimensional integral for any $m \geq 1$. Thus, if we implement the schemes by QMC method, the integration error of NV scheme is larger than that of the proposed scheme.
- The numerical algorithm of the NV scheme depends on the solvability of an ordinary differential equation (ODE) $dx_t/dt = V_i(x_t)$ with the vector fields $V_i = \sigma_i$, $i = 1, \ldots, d$ and $V_0 = \sigma_0 + (1/2)\sum_{i=1}^d \mathcal{L}_i \sigma_i$. This is because the NV scheme requires to compose ODE flows $\exp(tV_0)(\cdot)$ or random ODE flows $\exp(tZ^i V_i)(\cdot)$, $i = 1, \ldots, d$, $Z^i \sim N(0, 1)$ in (5.21). When the vector fields are very simple or of specific types, ODE flows can be analytically solved (the case is called "lucky case" in Bayer et al. (2013) and Morimoto and Sasada (2017) and hence, NV scheme is computed faster in such a case. The problem occurs when the ODE flows $\exp(tV_i)(\cdot)$, $i = 0, 1, \ldots, d$ do not have analytical solutions. In the case that ODE flows has no analytical solutions, The NV scheme has to rely on some numerical method for solving the ODE, which may cause time-consuming computation.[1] Thus, in the NV scheme, the form of the vector fields and corresponding ODEs should be checked and the numerical method must be chosen according to the target

---

[1] Here, note that Shinozaki (2017) introduced a third order method in this direction and a semi-closed form approximation scheme was proposed by Yamada (2022).

Table 5.1 The proposed scheme and Ninomiya-Victoir scheme

Method	m-order Weak Approx	Ninomiya-Victoir
Discretization error	$O\left(\frac{1}{n^m}\right)$ $(m \geq 2)$	$O\left(\frac{1}{n^2}\right)$
Integration error	$O\left(\frac{(\log M)^{n \times d}}{M}\right)$	$O\left(\frac{(\log M)^{n \times (d+1)}}{M}\right)$
ODE flows $e^{sV_i}(\cdot)$	Do not need analytical tractability	Need analytical or numerical solution

models. On the contrary, the proposed scheme works whether ODE flows can be solved analytically or not, because it only needs the Euler-Marumaya scheme and the Malliavin weight $\mathcal{M}^{(m)}$.

The features of the proposed scheme and NV scheme are summarized in the Table 5.1.

5.6 Notes and Summary

This chapter showed a m-order weak approximation of SDE:

$$\left|\mathbb{E}[f(X_T^x)] - \mathbb{E}[f(\bar{X}_T^{\text{EM},x,(n)}) \prod_{i=1}^{n} \mathcal{M}^{(m)}(T/n, \bar{X}_{(i-1)T/n}^{\text{EM},x,(n)}, W_{iT/n} - W_{(i-1)T/n})]\right|$$

$$= \|f\|_{\infty} \times O\left(\frac{1}{n^m}\right)$$

for a bounded measurable function f, based on Yamada (2019) which introduced a weak approximation scheme with a generalized Malliavin weight by extending the asymptotic expansion-based discretization method of Takahashi and Yamada (2016).

The computational schemes are studied in Naito and Yamada (2019), Yamada and Yamamoto (2020), Iguchi and Yamada (2021, 2022) and Yamada (2021, 2023). The principle can be applied to implementation of the Milstein scheme, multilevel Monte-Carlo method, automatic differentiation, nonlinear problems and statistical inference. See Yamada and Yamamoto (2018), Yamada and Yamamoto (2019), Naito and Yamada (2019), Naito and Yamada (2022a), Iguchi et al. (2024a,b), Iguchi and Beskos (2025), Iguchi et al. (2025) and Yamada (2018, 2023) for example.

Chapter 6
Application: Deep Learning-Based Weak Approximation

In the chapter, using deep learning effectively we apply the weak approximation method introduced in the previous chapter for solving a parabolic partial differential equation (PDE) and a backward dynamic programming problem. More concretely, we will solve nested conditional expectations appearing in Bermudan option pricing problem by deep learning-based least squares regression combined with the weak approximation under high-dimensional diffusion settings.

6.1 Introduction: Weak Approximation and Partial Differential Equation

In Chaps. 3 and 4, we have introduced the approaches of asymptotic expansion and weak approximation. Since a map given by an expectation of a solution of SDE corresponds to a solution to a parabolic PDE by Feynman-Kac formula (Theorem 2.15), the asymptotic expansion and the weak approximation may be applied to solve PDEs. On the other hand, numerical schemes such as Monte-Carlo method for the asymptotic expansion and the weak approximation give approximation of $\mathbb{E}[f(X_T^x)]$ for a fixed $x \in \mathbb{R}^N$, not approximation of a map $x \mapsto \mathbb{E}[f(X_T^x)]$, where $\{X_t^x\}_t$ is a solution of a SDE starting from x.

Deep learning solves this problem by minimizing a certain loss function, which means that a deep learning-based finite element method can be constructed for solving PDEs.

We briefly summarize the principle. Let $(\Omega, \mathcal{F}, P)$ be a probability space. Let $\{X_x\}_{x \in [a,b]^d} : [a, b]^d \times \Omega \to \mathbb{R}$ be a measurable map such that $X_x \in L^2(\Omega)$ for all $x \in [a, b]^d$, and $x \mapsto \mathbb{E}[X_x]$ is continuous. Then,

$$\int_{[a,b]^d} \mathbb{E}[|X_x - \mathbb{E}[X_x]|^2]dx = \inf_{v \in C([a,b]^d)} \int_{[a,b]^d} \mathbb{E}[|X_x - v(x)|^2]dx.$$

See Beck et al. (2021a) for instance. The fact suggests that the map $x \mapsto \mathbb{E}[X_x]$ is represented by a minimization problem. By applying the principle, we can extend the asymptotic expansion and weak approximation in Chaps. 3 and 4 to a type of spatial approximations.

Let $(\Omega, \mathcal{F}, P)$ be a probability space on which a Brownian motion is defined. Let X be the solution of following SDE driven by a d-dimensional Brownian motion W:

$$dX_t^x = \sigma_0(X_t^x)dt + \sum_{i=1}^{d} \sigma_i(X_t^x)dW_t^i, \quad X_0^x = x \in \mathbb{R}^N, \tag{6.1}$$

where $\sigma_i \in C_b^\infty(\mathbb{R}^N; \mathbb{R}^N)$, $i = 0, 1, \ldots, d$ and assume the uniformly elliptic condition. Let $f \in C(\mathbb{R}^N)$ be a bounded function, and let $u \in C([0, T] \times \mathbb{R}^N) \cap C^{1,2}([0, T) \times \mathbb{R}^N)$ be a solution of the following PDE:

$$(\partial_t + \mathcal{L})u(t, x) = 0, \quad (t, x) \in [0, T) \times \mathbb{R}^N,$$
$$u(T, x) = f(x), \quad x \in \mathbb{R}^N,$$

where $\mathcal{L}$ is the corresponding generator of the SDE. Then, by the Feynman-Kac formula (Theorem 2.15), we have

$$u(t, x) = P_{T-t} f(x) = \mathbb{E}[f(X_{T-t}^x)], \quad t \in [0, T], \quad x \in \mathbb{R}^N.$$

For $t \in [0, T)$, the map $x \mapsto u(t, x)$ can be approximated using the m-order weak approximation scheme in the following way. Let $\{Q_s^{(m)}\}_{s>0}$ be the local approximation operator, i.e.

$$Q_s^{(m)}\varphi(x) = \mathbb{E}[\varphi(\bar{X}_s^{\mathrm{EM},x})\mathcal{M}^{(m)}(s, x, W_s)], \quad s > 0, \quad x \in \mathbb{R}^N, \quad \varphi \in \mathcal{B}_b(\mathbb{R}^N),$$

where $\bar{X}_s^{\mathrm{EM},x}$ is the one-step Euler-Maruyama scheme and $\mathcal{M}^{(m)}$ is the Malliavin weight function introduced in (5.18) in the previous chapter. Using the principle above, the m-order weak approximation function on a cube $[a, b]^N$ is represented by the following minimization problem:

$$[[a, b]^N \ni x \mapsto (Q_{(T-t)/n}^{(m)})^n f(x)]$$
$$= \mathrm{argmin}_{v \in C([a,b]^N)} \mathbb{E}[|v(\xi)$$
$$- f(\bar{X}_{T-t}^{\mathrm{EM},\xi,(n)}) \prod_{i=1}^{n} \mathcal{M}^{(m)}((T-t)/n, \bar{X}_{(i-1)(T-t)/n}^{\mathrm{EM},\xi,(n)}, W_{i(T-t)/n} - W_{(i-1)(T-t)/n})|^2],$$

where $\{\bar{X}_{k(T-t)/n}^{\mathrm{EM},x,(n)}\}_{k \leq n}$ is the Euler-Maruyama scheme starting from $x \in \mathbb{R}^N$ and $\xi : \Omega \to [a, b]^N$ is an independent uniformly distributed random variable, i.e. $\xi \sim U([a, b]^N)$. Here, the weak approximation function $[a, b]^N \ni x \mapsto (Q_{(T-t)/n}^{(m)})^n f(x)$ can be arbitrarily approximated by a neural network

$[a,b]^N \ni x \mapsto [(Q_{(T-t)/n}^{(m)})^n f]^{\mathcal{NN}}(x)$ using the above minimization problem as a loss function through the fact that a space of neural networks is a dense subset of $C([a,b]^N)$. As a consequence, the PDE solution map $[a,b]^N \ni x \mapsto u(t,x)$ can be discretized by

$$u(t,\cdot) = [(Q_{(T-t)/n}^{(m)})^n f]^{\mathcal{NN}}(\cdot) + O(n^{-m}).$$

Therefore, the scheme will be a high order weak approximation-based finite element method for solving PDEs. Iguchi et al. (2021), Naito and Yamada (2022a) provided theoretical and numerical results for such a scheme. Moreover, Takahashi and Yamada (2023) showed that an asymptotic expansion with deep learning approximates a solution of a parabolic PDE without suffering the curse of dimensionality.

Based on the method, we apply the weak approximation scheme of the previous chapter to an important nonlinear problem called backward dynamic programming principle.

6.2 Backward Dynamic Programming Principle

Let $(\Omega, \mathcal{F}, \mathbb{P})$ be a probability space with a filtration $\{\mathcal{F}_t\}_{t\geq 0}$ and let X be an adapted process. Consider a problem of determining the most suitable time $\tau \in \{t_0, t_1, \ldots, t_K\}$ in the game that the player receives a payoff $f(t_i, X_{t_i})$ at time t_i. The expected value Y_{t_i} at time t_i is the maximum between $f(t_i, X_{t_i})$ and the expected value at t_{i+1} expected from t_i, that is

$$Y_{t_i} = \max\{f(t_i, X_{t_i}), \mathbb{E}[Y_{t_{i+1}}|\mathcal{F}_{t_i}]\}, \quad i = 0, 1, \ldots, K-1 \quad (6.2)$$

with the terminal $Y_{t_K} = f(t_K, X_{t_K})$. Moreover, if X is a Markov process, a optimal stopping problem:

$$v_{t_i}(\cdot) = \sup_{\tau \in \mathcal{T}_i} \mathbb{E}[f(\tau, X_\tau)|X_{t_i} = \cdot], \quad i = 0, 1, \ldots, K,$$

where $\mathcal{T}_i = \{\tau : \Omega \to \{t_i, \ldots, t_K\}; \tau \text{ is stopping time}\}$ for $i = 0, 1, \ldots, K$, is characterized by a backward dynamic programming principle as follows:

$$v_{t_i}(\cdot) = \max\{f(t_i, \cdot), \mathbb{E}[v_{t_{i+1}}(X_{t_{i+1}})|X_{t_i} = \cdot]\}, \quad i = 0, 1, \ldots, K-1 \quad (6.3)$$

with the terminal $v_{t_K}(\cdot) = f(t_K, \cdot)$. See Sect. 7.4 of Gobet (2020) and Sect. 11.2 in Pages (2018) for the details.

We formulate a backward dynamic programming principle in a general multi-dimensional diffusion setting as in Naito and Yamada (2024b). Hereafter, in this chapter, let $(\Omega, \mathcal{F}, P)$ be a probability space on which a d-dimensional Brownian motion $W = \{W_t\}_t$ is defined.

Let us consider the solution of the following N-dimensional SDE driven by a d-dimensional Brownian motion:

$$X_s^{t,x} = x + \int_t^s \sigma_0(X_r^{t,x})dr + \sum_{i=1}^d \int_t^s \sigma_i(X_r^{t,x})dW_r^i, \quad X_t^{t,x} = x \in \mathbb{R}^N, \quad (6.4)$$

where $\sigma_i \in C_b^\infty(\mathbb{R}^N; \mathbb{R}^N)$, $i = 0, 1, \ldots, d$. We assume the uniformly elliptic condition for the matrix $\sigma = (\sigma_1, \ldots, \sigma_d)$.

Let $K \in \mathbb{N}$ be a fixed natural number of the early-exercise dates of a Bermudan option and let τ_j, $j = 1, \ldots, K$ denote the early-exercise dates given by

$$0 =: \tau_0 < \tau_1 < \cdots < \tau_K = T$$

and $\tau_K - \tau_{K-1} = \cdots = \tau_1 - \tau_0$. For numerical computation, we choose the number of time steps $n \in \mathbb{N}$ for each interval of the early-exercise dates, i.e.

$$0 = t = t_0 < t_1 < \cdots < t_{n-1} < t_n < t_{n+1} < \cdots < t_{nK-1} < t_{nK} = T,$$
$$\tau_j = t_{jn}, \quad t_{jn} - t_{(j-1)n} = T/K, \quad j = 1, \ldots, K,$$
$$t_i - t_{i-1} = T/(nK), \quad i = 1, \ldots, nK,$$

or $t_{jn+\ell} = \tau_j + (\tau_{j+1} - \tau_j)\ell/n$, $j = 0, 1, \ldots, K-1$, $\ell = 0, 1, \ldots, n$.

Let $f : \{\tau_0, \ldots, \tau_K\} \times \mathbb{R}^N \to \mathbb{R}$ be a bounded measurable function. For $j = 0, 1, \ldots, K$, $f(\tau_j, \cdot)$ represents the discounted payoff function at the early-exercise date τ_j. Then, the value functions of the Bermudan option written on the underlying asset X on the early-exercise dates is given by:

$$v_{\tau_K}(x) = f(\tau_K, x), \quad (6.5)$$
$$v_{\tau_j}(x) = (P_{\tau_j, \tau_{j+1}} v_{\tau_{j+1}})(x) \vee f(\tau_j, x), \quad j = K-1, \ldots, 1, 0, \quad (6.6)$$

for $x \in \mathbb{R}^N$, where $a \vee b = \max\{a, b\}$ and

$$P_{\tau_j, \tau_{j+1}} g(x) = \mathbb{E}[g(X_{\tau_{j+1}}^{\tau_j, x})], \quad x \in \mathbb{R}^N, \quad j = 0, 1, \ldots, K-1, \quad (6.7)$$

for a bounded measurable function $g : \mathbb{R}^N \to \mathbb{R}$. In this chapter, we show a fast, efficient computation scheme for v_0 using the weak approximation with deep learning based on Naito and Yamada (2024a).

6.3 Sketch of Weak Approximation Scheme

The deep learning-based weak approximation scheme and the algorithm are briefly summarized as follows:

1. For each j-th date of early-exercise dates $0 < \tau_1 < \cdots < \tau_K = T$, we discretize the interval $[\tau_j, \tau_{j+1}]$ n-times and (given Θ^*_{j+1}) estimate the optimal parameter Θ^*_j for a deep neural network backward from the terminal date:

$$\Theta^*_j = \mathrm{argmin}_{\Theta \in \mathbb{R}^\nu} \mathbb{E}\big[|\mathbf{DeepNN}^{\Theta}_j(\bar{X}^n_{\tau_j})$$
$$- (\mathbf{DeepNN}^{\Theta^*_{j+1}}_{j+1} \vee f(\tau_{j+1}, \cdot))(\bar{X}^n_{\tau_{j+1}}) \times \mathbf{Malliavin\ weight}^{(m),n}_{j,j+1}|^2\big]$$

where $\mathbf{DeepNN}^{\Theta}_j$ is a deep neural network (which will be described later) with a set of parameters Θ at j-th date, $f(\tau_j, \cdot)$ is the discounted payoff function at j-th date, $\bar{X}^n_{\tau_j}$ is the Euler-Maruyama scheme at j-th date which starts a fixed $x \in \mathbb{R}^q$ at $\tau_0 = 0$, and $\mathbf{Malliavin\ weight}^{(m),n}_{j,j+1}$ is the m-order Malliavin weight on the interval. If $j = K - 1$, the algorithm is replaced with

$$\Theta^*_{K-1} = \mathrm{argmin}_{\Theta \in \mathbb{R}^\nu} \mathbb{E}[|\mathbf{DeepNN}^{\Theta}_{K-1}(\bar{X}^n_{\tau_{K-1}}) - f(\tau_K, \bar{X}^n_{\tau_K}) \times \mathbf{Malliavin\ weight}^{(m),n}_{K-1,K}|^2].$$

2. On the interval $[0, \tau_1]$, we perform simple (single) parameter estimation:

$$\theta^* = \mathrm{argmin}_{\theta \in \mathbb{R}} \mathbb{E}[|\theta - \mathbf{DeepNN}^{\Theta^*_1}_1(\bar{X}^n_{\tau_1}) \times \mathbf{Malliavin\ weight}^{(m),n}_{0,1}|^2]$$

and we finally obtain an accurate approximation $v^{(m),n\,*}_0 = \theta^* \vee f(\tau_0, x)$ such that

$$|v_0 - v^{(m),n\,*}_0| = O(n^{-m}).$$

The advantages of the scheme are summarized as follows:

- The scheme gives a fast, efficient computation for the value functions on early-exercise dates without the curse of dimensionality due to the advantages of higher-order discretization and deep learning, which will be an extension of the methods of Carriere (1996), Longstaff and Schwartz (2001), Tsitsiklis and Van (2001);
- The scheme with the m-order Malliavin weight accelerates the convergence of the approximation as $O(n^{-m})$, which is justified by a weak approximation theory, and the algorithm is constructed for arbitrary $m \geq 1$;
- The m-order Malliavin weight is always computed with simulation of increments of d-dimensional Brownian motion for all $m \geq 1$, in other words, the cost of generating random numbers is the same as that of the Euler-Maruyama scheme.

6.4 Deep Neural Network and Universal Approximation

We briefly introduce the deep neural network from mathematical perspective. For $r \in \mathbb{N}$, let $\mathcal{L}_r : \mathbb{R}^r \to \mathbb{R}^r$ be an activation function given by $\mathbb{R}^r \ni x \mapsto \mathcal{L}_r(x) = (\mathcal{L}(x_1), \ldots, \mathcal{L}(x_r))$ with a nonlinear function $\mathcal{L} : \mathbb{R} \to \mathbb{R}$. For $p, \ell \in \mathbb{N}, e \in \{0\} \cup \mathbb{N}$ and $\theta = (\theta^1, \ldots, \theta^\nu) \in \mathbb{R}^\nu$ such that $e + \ell(p+1) \leq \nu$, let $A_{p,\ell}^{\theta,e} : \mathbb{R}^p \to \mathbb{R}^\ell$ be an affine transformation function given by

$$A_{p,\ell}^{\theta,e}(x) = \begin{pmatrix} \theta^{e+1} & \cdots & \theta^{e+p} \\ \vdots & \ddots & \vdots \\ \theta^{e+(\ell-1)p+1} & \cdots & \theta^{e+\ell p} \end{pmatrix} \begin{pmatrix} x_1 \\ \vdots \\ x_p \end{pmatrix} + \begin{pmatrix} \theta^{e+\ell p+1} \\ \vdots \\ \theta^{e+\ell p+\ell} \end{pmatrix}, \quad x \in \mathbb{R}^p.$$

Let $s \in \{3, 4, 5, \ldots\}$ be a fixed integer. Let $d_0, d_1, \ldots, d_s \in \mathbb{N}$ and $\nu = \sum_{i=1}^s d_i(d_{i-1}+1)$. Then, we define a deep neural network $\Phi : \mathbb{R}^{d_0} \to \mathbb{R}^{d_s}$ as

$$\Phi(x) = (A_{d_{s-1},d_s}^{\theta, \sum_{i=1}^{s-1} d_i(d_{i-1}+1)} \circ \mathcal{L}_{d_{s-1}} \circ A_{d_{s-2},d_{s-1}}^{\theta, \sum_{i=1}^{s-2} d_i(d_{i-1}+1)}$$
$$\circ \cdots \circ \mathcal{L}_{d_2} \circ A_{d_1,d_2}^{\theta, d_1(d_0+1)} \circ \mathcal{L}_{d_1} \circ A_{d_0,d_1}^{\theta,0})(x), \quad x \in \mathbb{R}^{d_0}. \quad (6.8)$$

Here, d_i represents the number of neurons of of i-th layer for $i = 0, 1, \ldots, s$, and ν is the number of parameters of the deep neural network. A typical choice of activation function is the ReLU (Rectified Linear Unit) function given by $\mathcal{L}_r^{\text{ReLU}}(x) = ((x_1)^+, \ldots, (x_r)^+), x \in \mathbb{R}^r$ for $r \in \mathbb{N}$, where $(\cdot)^+ := \max\{\cdot, 0\}$.

Here, we summarize a result so that the conditional expectations appearing in the backward dynamic programming are appropriately approximated by ReLU neural networks as a version of universal approximation theorem (UAT) which may be regarded as an extension of those in Hornik et al. (1989, 1990), Hornik (1991), Shigekawa (2004) and Calin (2020).

Theorem 6.1 (UAT) *Let $(\Omega, \mathcal{F}, P)$ be a probability space. Let $X : \Omega \to \mathbb{R}^N$ and $Y : \Omega \to \mathbb{R}^N$ be square-integrable random variables. Then, the map $\mathbb{E}[Y|X = \cdot]$ can be arbitrarily approximated by a ReLU neural network, i.e. for all $\varepsilon > 0$, there exists a ReLU neural network $\varphi_\varepsilon^{\mathcal{NN}}$ with at most $L = 2(\lfloor \log_2(N+1) \rfloor + 2)$ layers such that $\|\mathbb{E}[Y|X = \cdot] - \varphi_\varepsilon^{\mathcal{NN}}\|_{L^2(\mathbb{R}^N, \mathcal{B}(\mathbb{R}^N), P \circ X^{-1})} < \varepsilon$.*

Proof of Theorem 6.1. It holds that

$$\mathbb{E}[|Y - \mathbb{E}[Y|X]|^2] = \inf_{Z \in L^2(\Omega, \sigma(X), P)} \mathbb{E}[|Y - Z|^2] \quad (6.9)$$

by (2.1. Since for a $\sigma(X)$-measurable function Z, there exists a Borel measurable function $g : \mathbb{R}^N \to \mathbb{R}$ such that $Z = g(X)$ by Theorem 2.1, and $g(X) \in L^2(\Omega, \sigma(X), P)$ (i.e. $|g(X)|^2 \in L^1(\Omega, \sigma(X), P)$) is equivalent to $g \in L^2(\mathbb{R}^N, \mathcal{B}(\mathbb{R}^N), P \circ X^{-1})$ (i.e. $|g|^2 \in L^1(\mathbb{R}^N, \mathcal{B}(\mathbb{R}^N), P \circ X^{-1})$) by Theorem 2.2, (6.9) can be expressed by

6.4 Deep Neural Network and Universal Approximation

$$\mathbb{E}[|Y - g^*(X)|^2] = \inf_{g \in L^2(\mathbb{R}^N, \mathscr{B}(\mathbb{R}^N), P \circ X^{-1})} \mathbb{E}[|Y - g(X)|^2]$$

where $g^*(\cdot) = \mathbb{E}[Y|X = \cdot] \in L^2(\mathbb{R}^N, \mathscr{B}(\mathbb{R}^N), P \circ X^{-1})$.

Let $C_0(\mathbb{R}^N)$ be the space of compact supported continuous functions on $\mathbb{R}^N$ and let $C_{pl}(\mathbb{R}^N)$ be the space of continuous, piecewise linear functions on $\mathbb{R}^N$. We note that $C_{pl}(\mathbb{R}^N)$ is a dense subset of $C_0(\mathbb{R}^N)$.

Since the measure $P \circ X^{-1}$ is regular by Theorem 2.18 of Rudin (1987), we can see that by Theorem 3.14 of Rudin (1987), $C_0(\mathbb{R}^N) \subset L^2(\mathbb{R}^N, \mathscr{B}(\mathbb{R}^N), P \circ X^{-1})$ is dense. Let $\varepsilon > 0$ be a fixed small real number. Then, $\exists \varphi_1 \in C_0(\mathbb{R}^N)$ s.t. $\|g^* - \varphi_1\|_{L^2(\mathbb{R}^N, \mathscr{B}(\mathbb{R}^N), P \circ X^{-1})} < \varepsilon/2$. Also, $\exists \varphi_2 \in C_{pl}(\mathbb{R}^N)$ s.t. $\|\varphi_1 - \varphi_2\|_{L^2(\mathbb{R}^N, \mathscr{B}(\mathbb{R}^N), P \circ X^{-1})} < \varepsilon/2$. Note that φ_2 can be represented as a linear combination of piecewise linear convex functions as $\varphi_2 = \sum_{j=1}^{q} s_j g_j$ where $s_j \in \{-1, 1\}$, $j \leq q$ and g_j, $j \leq q$ are piecewise linear convex functions having at most $N + 1$ pieces (Theorem 1 in Wang and Sun 2005), whose values are computed by a ReLU neural network with at most $2(\lfloor \log_2(N + 1) \rfloor + 1)$ hidden layers[1] based on Proposition 10.2.2 in Calin (2020). Then φ_2 is represented by $\varphi_2 = \varphi_\varepsilon^{NN}$ with a ReLU neural network φ_ε^{NN} (with one input layer and one output layer) with at most $L = 1 + 2(\lfloor \log_2(N + 1) \rfloor + 1) + 1 = 2(\lfloor \log_2(N + 1) \rfloor + 2)$ layers. Therefore,

$$\|g^* - \varphi_\varepsilon^{NN}\|_{L^2(\mathbb{R}^N, \mathscr{B}(\mathbb{R}^N), P \circ X^{-1})}$$
$$< \|g^* - \varphi_1\|_{L^2(\mathbb{R}^N, \mathscr{B}(\mathbb{R}^N), P \circ X^{-1})} + \|\varphi_1 - \varphi_2\|_{L^2(\mathbb{R}^N, \mathscr{B}(\mathbb{R}^N), P \circ X^{-1})}$$
$$< \varepsilon/2 + \varepsilon/2 = \varepsilon. \quad \square$$

The following result is useful in practice, which will be used in the next section.

Corollary 6.1 *Let $(\Omega, \mathcal{F}, P)$ be a probability space. Let $X : \Omega \to \mathbb{R}^N$ and $Y : \Omega \to \mathbb{R}^N$ be random variables, which are independent. Let $f : \mathbb{R}^N \times \mathbb{R}^N \to \mathbb{R}$ be a Borel measurable function and suppose that $f(Y, X) \in L^2(\Omega, \mathcal{F}, P)$. Then, the map $\mathbb{E}[f(Y, X)|X = \cdot] = \mathbb{E}[f(Y, \cdot)] \in L^2(\mathbb{R}^N, \mathscr{B}(\mathbb{R}^N), P \circ X^{-1})$ can be arbitrarily approximated by a ReLU neural network with at most $L = 2(\lfloor \log_2(N + 1) \rfloor + 2)$ layers.*

Proof of Corollary 6.1. By the assumption, it holds that $\mathbb{E}[f(Y, X)|X] = \mathbb{E}[f(Y, x)]|_{x=X}$ and

$$\mathbb{E}[|f(Y, X) - g^*(X)|^2] = \inf_{g \in L^2(\mathbb{R}^N, \mathscr{B}(\mathbb{R}^N), P \circ X^{-1})} \mathbb{E}[|f(Y, X) - g(X)|^2],$$

where $g^*(\cdot) = \mathbb{E}[f(Y, X)|X = \cdot] = \mathbb{E}[f(Y, \cdot)] \in L^2(\mathbb{R}^N, \mathscr{B}(\mathbb{R}^N), P \circ X^{-1})$. Then, Theorem 6.1 gives the assertion. $\square$

---

[1] We thank Prof. Calin for a comment on this issue.

6.5 Weak Approximation with Deep Learning

We now show a weak approximation scheme for the backward dynamic programming principle summarized in Sect. 6.3, using the method in Chap. 5 with the result of Sect. 6.4.

For $i = jn$, $j = 0, 1, \ldots, K-1$, let $\bar{X}_{t_\ell}^{\mathrm{EM}, t_i, x}$, $\ell \geq i$ be the Euler-Maruyama scheme starting at t_i:

$$\bar{X}_{t_i}^{\mathrm{EM}, t_i, x} = x \in \mathbb{R}^N,$$

$$\bar{X}_{t_\ell}^{\mathrm{EM}, t_i, x} = \bar{X}_{t_{\ell-1}}^{\mathrm{EM}, t_i, x} + \sigma_0(\bar{X}_{t_{\ell-1}}^{\mathrm{EM}, t_i, x})(t_\ell - t_{\ell-1}) + \sum_{j=1}^d \sigma_j(\bar{X}_{t_{\ell-1}}^{\mathrm{EM}, t_i, x})(W_{t_\ell}^j - W_{t_{\ell-1}}^j), \; \ell > i.$$

(6.10)

Let $m \in \mathbb{N}$. We define

$$Q_{t_i, t_{i+1}}^{(m)} g(x) = \mathbb{E}[g(\bar{X}_{t_{i+1}}^{\mathrm{EM}, t_i, x}) \mathcal{W}_{t_{i+1}}^{(m), t_i, x}],$$

$$Q_{\tau_j, \tau_{j+1}}^{(m), n} g(x) = Q_{\tau_j, t_{jn+1}}^{(m)} \cdots Q_{t_{(j+1)n-1}, \tau_{j+1}}^{(m)} g(x) = \mathbb{E}\left[g(\bar{X}_{\tau_{j+1}}^{\mathrm{EM}, \tau_j, x}) \prod_{\ell=1}^n \mathcal{W}_{t_{jn+\ell}}^{(m), t_{jn+\ell-1}, \bar{X}_{t_{jn+\ell-1}}^{\mathrm{EM}, \tau_j, x}} \right]$$

(6.11)

for a bounded measurable function $g : \mathbb{R}^N \to \mathbb{R}$, where $\mathcal{W}_{t_{i+1}}^{(m), t_i, x}$ is the Malliavin weight given by

$$\mathcal{W}_{t_{i+1}}^{(m), t_i, x} = \mathcal{M}^{(m)}(t_{i+1} - t_i, x, W_{t_{i+1}} - W_{t_i}).$$

(6.12)

Using the results of Chap. 4 and Sect. 6.4, we have the followings.

Lemma 6.1 *There exists $C > 0$ independent of K such that*

$$\|P_{\tau_j, \tau_{j+1}} g - Q_{\tau_j, \tau_{j+1}}^{(m), n} g\|_\infty \leq C \|g\|_\infty K^{-1} n^{-m},$$

(6.13)

for all bounded measurable functions $g : \mathbb{R}^N \to \mathbb{R}$, $j = 0, 1, \ldots, K-1$ and $n \geq 1$.

Lemma 6.2 *Let $g : \mathbb{R}^N \to \mathbb{R}$ be a bounded measurable function and let $0 \leq i < j \leq K-1$. For $n \in \mathbb{N}$, there exists a neural network $Q_{\tau_j, \tau_{j+1}}^{(m), n, \mathcal{NN}} g$ such that*

$$\|Q_{\tau_j, \tau_{j+1}}^{(m), n} g - Q_{\tau_j, \tau_{j+1}}^{(m), n, \mathcal{NN}} g\|_{L^2(\mathbb{R}^N, \mathcal{B}(\mathbb{R}^N), P \circ (\bar{X}_{\tau_j}^{\mathrm{EM}, \tau_i, x})^{-1})} \leq K^{-1} n^{-m}.$$

(6.14)

6.5 Weak Approximation with Deep Learning

Proof The map $Q^{(m),n}_{\tau_j,\tau_{j+1}} g$ has the form:

$$Q^{(m),n}_{\tau_j,\tau_{j+1}} g = \mathbb{E}\left[g(\bar{X}^{EM,\tau_i,x}_{\tau_{j+1}}) \prod_{\ell=1}^{n} \mathcal{W}^{(m),t_{jn+\ell-1},\bar{X}^{EM,\tau_i,x}_{t_{jn+\ell-1}}}_{t_{jn+\ell}} \Big| \bar{X}^{EM,\tau_i,x}_{\tau_j} = \cdot \right] \quad (6.15)$$

where $t_{jn+\ell} = \tau_j + (\tau_{j+1} - \tau_j)\ell/n$, $\ell = 0, 1, \ldots, n$, and is approximated by UAT in Theorem 6.1 or Corollary 6.1 as follows: for all $\varepsilon > 0$, there is a ReLU neural network $\psi_{\varepsilon,g,j,n}$ such that

$$\| Q^{(m),n}_{\tau_j,\tau_{j+1}} g - \psi_{\varepsilon,g,j,m,n} \|_{L^2(\mathbb{R}^N,\mathcal{B}(\mathbb{R}^N),P\circ(\bar{X}^{EM,\tau_i,x}_{\tau_j})^{-1})} < \varepsilon. \quad (6.16)$$

Define $Q^{(m),n,\mathcal{NN}}_{\tau_j,\tau_{j+1}} g := \psi_{K^{-1}n^{-m},g,j,m,n}$. $\square$

We define the following functions based on the neural networks: for $x \in \mathbb{R}^N$,

$$\begin{aligned}
\bar{v}^{(m),n,\mathcal{NN}}_{\tau_K}(x) &= f(\tau_K, x), \\
\bar{v}^{(m),n,\mathcal{NN}}_{\tau_j}(x) &= (Q^{(m),n,\mathcal{NN}}_{\tau_j,\tau_{j+1}} \bar{v}^{(m),n,\mathcal{NN}}_{\tau_{j+1}})(x) \vee f(\tau_j, x), \quad j = K-1, \ldots, 1, \\
\bar{v}^{(m),n,\mathcal{NN}}_0(x) &= (Q^{(m),n}_{\tau_0,\tau_1} \bar{v}^{(m),n,\mathcal{NN}}_{\tau_1})(x) \vee f(\tau_0, x).
\end{aligned} \quad (6.17)$$

Then, the following holds.

Theorem 6.2 (*m-order weak approximation for Bermudan option price v_0*) *There exists $C > 0$ such that*

$$\left| v_0(x) - \bar{v}^{(m),n,\mathcal{NN}}_0(x) \right| \leq C \frac{1}{n^m} \quad (6.18)$$

for all $n \geq 1$.

Proof Using the inequality $|x \vee z - y \vee z| \leq |x - y|$, we have

$$\begin{aligned}
& |v_0(x) - \bar{v}^{(m),n,\mathcal{NN}}_0(x)| \\
& \leq |P_{\tau_0,\tau_1} v_{\tau_1}(x) - Q^{(m),n}_{\tau_0,\tau_1} \bar{v}^{(m),n,\mathcal{NN}}_{\tau_1}(x)| \\
& \leq |(P_{\tau_0,\tau_1} - Q^{(m),n}_{\tau_0,\tau_1}) v_{\tau_1}(x)| + |Q^{(m),n}_{\tau_0,\tau_1}(v_{\tau_1} - \bar{v}^{(m),n,\mathcal{NN}}_{\tau_1})(x)|.
\end{aligned} \quad (6.19)$$

By Lemma 6.1,

$$|(P_{\tau_0,\tau_1} - Q^{(m),n}_{\tau_0,\tau_1}) v_{\tau_1}(x)| \leq C \| v_{\tau_1} \|_\infty K^{-1} n^{-m}, \quad (6.20)$$

and by the definition of $Q_{\tau_0,\tau_1}^{(m),n}$,

$$|Q_{\tau_0,\tau_1}^{(m),n}(v_{\tau_1} - \bar{v}_{\tau_1}^{(m),n,\mathcal{NN}})(x)|$$
$$\leq \|v_{\tau_1} - \bar{v}_{\tau_1}^{(m),n,\mathcal{NN}}\|_{L^2(\mathbb{R}^N,\mathscr{B}(\mathbb{R}^N),P\circ(\bar{X}_{\tau_1}^{EM,\tau_0,x})^{-1})} \Big\| \prod_{\ell=1}^{n} W_{t_\ell}^{(m),t_{\ell-1},\bar{X}_{t_{\ell-1}}^{EM,\tau_0,x}} \Big\|_{L^2(\Omega)}$$
$$\leq \|v_{\tau_1} - \bar{v}_{\tau_1}^{(m),n,\mathcal{NN}}\|_{L^2(\mathbb{R}^N,\mathscr{B}(\mathbb{R}^N),P\circ(\bar{X}_{\tau_1}^{EM,\tau_0,x})^{-1})}(1+c/(nK))^n$$
$$\leq \|v_{\tau_1} - \bar{v}_{\tau_1}^{(m),n,\mathcal{NN}}\|_{L^2(\mathbb{R}^N,\mathscr{B}(\mathbb{R}^N),P\circ(\bar{X}_{\tau_1}^{EM,\tau_0,x})^{-1})} e^{c/K}. \quad (6.21)$$

For $j = 1, \ldots, K-2$, we have

$$\|v_{\tau_j} - \bar{v}_{\tau_j}^{(m),n,\mathcal{NN}}\|_{L^2(\mathbb{R}^N,\mathscr{B}(\mathbb{R}^N),P\circ(\bar{X}_{\tau_j}^{EM,\tau_{j-1},x})^{-1})}$$
$$\leq \|(P_{\tau_j,\tau_{j+1}} - Q_{\tau_j,\tau_{j+1}}^{(m),n,\mathcal{NN}})v_{\tau_{j+1}}\|_{L^2(\mathbb{R}^N,\mathscr{B}(\mathbb{R}^N),P\circ(\bar{X}_{\tau_j}^{EM,\tau_{j-1},x})^{-1})}$$
$$+ \|Q_{\tau_j,\tau_{j+1}}^{(m),n,\mathcal{NN}}(v_{\tau_{j+1}} - \bar{v}_{\tau_{j+1}}^{(m),n,\mathcal{NN}})\|_{L^2(\mathbb{R}^N,\mathscr{B}(\mathbb{R}^N),P\circ(\bar{X}_{\tau_j}^{EM,\tau_{j-1},x})^{-1})}, \quad (6.22)$$

and

$$\|v_{\tau_{K-1}} - \bar{v}_{\tau_{K-1}}^{(m),n,\mathcal{NN}}\|_{L^2(\mathbb{R}^N,\mathscr{B}(\mathbb{R}^N),P\circ(\bar{X}_{\tau_{K-1}}^{EM,\tau_{K-2},x})^{-1})}$$
$$\leq \|(P_{\tau_{K-1},\tau_K} - Q_{\tau_{K-1},\tau_K}^{(m),n,\mathcal{NN}})v_{\tau_K}(x)\|_{L^2(\mathbb{R}^N,\mathscr{B}(\mathbb{R}^N),P\circ(\bar{X}_{\tau_{K-1}}^{EM,\tau_{K-2},x})^{-1})}, \quad (6.23)$$

since $v_{\tau_K} = \bar{v}_{\tau_K}^{(m),n,\mathcal{NN}} = f(\tau_K,\cdot)$. By the above lemmas, we have

$$\|P_{\tau_j,\tau_{j+1}}g - Q_{\tau_j,\tau_{j+1}}^{(m),n,\mathcal{NN}}g\|_{L^2(\mathbb{R}^N,\mathscr{B}(\mathbb{R}^N),P\circ(\bar{X}_{\tau_j}^{EM,\tau_{j-1},x})^{-1})} \leq C\|g\|_\infty K^{-1}n^{-m} + K^{-1}n^{-m}$$
$$(6.24)$$

and

$$\|Q_{\tau_j,\tau_{j+1}}^{(m),n,\mathcal{NN}}g\|_{L^2(\mathbb{R}^N,\mathscr{B}(\mathbb{R}^N),P\circ(\bar{X}_{\tau_j}^{EM,\tau_{j-1},x})^{-1})}$$
$$\leq \|Q_{\tau_j,\tau_{j+1}}^{(m),n}g\|_{L^2(\mathbb{R}^N,\mathscr{B}(\mathbb{R}^N),P\circ(\bar{X}_{\tau_j}^{EM,\tau_{j-1},x})^{-1})} + K^{-1}n^{-m}$$
$$\leq \Big(\int_{\mathbb{R}^N} \|g\|^2_{L^2(\mathbb{R}^N,\mathscr{B}(\mathbb{R}^N),P\circ(\bar{X}_{\tau_{j+1}}^{\tau_j,y})^{-1})} dP\circ(\bar{X}_{\tau_j}^{EM,\tau_{j-1},x})^{-1}(y)\Big)^{1/2}(1+c/(nK))^n + K^{-1}n^{-m}$$
$$\leq \Big(\int_{\mathbb{R}^N} \|g\|^2_{L^2(\mathbb{R}^N,\mathscr{B}(\mathbb{R}^N),P\circ(\bar{X}_{\tau_{j+1}}^{\tau_j,y})^{-1})} dP\circ(\bar{X}_{\tau_j}^{EM,\tau_{j-1},x})^{-1}(y)\Big)^{1/2} e^{c/K} + K^{-1}n^{-m} \quad (6.25)$$

for some $c, C > 0$ independent of g, j, K and n. Thus, we obtain

$$\|v_{\tau_j} - \bar{v}^{(m),n,\mathcal{NN}}_{\tau_j}\|_{L^2(\mathbb{R}^N, \mathscr{B}(\mathbb{R}^N), P \circ (\bar{X}^{\text{EM},\tau_{j-1},x}_{\tau_j})^{-1})}$$
$$\leq C\|v_{\tau_{j+1}}\|_\infty K^{-1} n^{-m}$$
$$+ e^{c/K} \left(\int_{\mathbb{R}^N} \|v_{\tau_{j+1}} - \bar{v}^{(m),n,\mathcal{NN}}_{\tau_{j+1}}\|^2_{L^2(\mathbb{R}^N, \mathscr{B}(\mathbb{R}^N), P \circ (\bar{X}^{\text{EM},\tau_j,y}_{\tau_{j+1}})^{-1})} \, dP \circ (\bar{X}^{\text{EM},\tau_{j-1},x}_{\tau_j})^{-1}(y) \right)^{1/2}$$
$$+ 2K^{-1} n^{-m}, \tag{6.26}$$

and

$$\|v_{\tau_{K-1}} - \bar{v}^{(m),n,\mathcal{NN}}_{\tau_{K-1}}\|_{L^2(\mathbb{R}^N, \mathscr{B}(\mathbb{R}^N), P \circ (\bar{X}^{\text{EM},\tau_{K-2},x}_{\tau_{K-1}})^{-1})} \leq C\|v_{\tau_K}\|_\infty K^{-1} n^{-m} + K^{-1} n^{-m}. \tag{6.27}$$

Therefore,

$$|v_0(x) - \bar{v}^{(m),n,\mathcal{NN}}_0(x)| \leq C \sum_{j=0}^{K-1} \|v_{\tau_{j+1}}\|_\infty K^{-1} n^{-m} + \sum_{j=1}^{K-1} 2K^{-1} n^{-m} = O\left(\frac{1}{n^m}\right) \tag{6.28}$$

where we used $\|v_{\tau_j}\|_\infty \leq \|v_{\tau_{j+1}}\|_\infty \vee \|f(\tau_j, \cdot)\|_\infty \leq \cdots \leq \max_{j \leq k \leq K} \|f(\tau_j, \cdot)\|_\infty$.
□

6.6 Algorithm, Implementation and Numerical Results

By the above discussion, one can construct a deep learning-based approximation $Q^{(m),n,\mathcal{NN},\theta^*_{\tau_j}}_{\tau_j,\tau_{j+1}} \bar{v}^{(m),n,\mathcal{NN}}_{\tau_{j+1}}$ parametrized by $\theta^*_{\tau_j} \in \mathbb{R}^\nu$ for $j = 0, 1, \ldots, n-1$ with a large enough $\nu \in \mathbb{N}$ as

$$Q^{(m),n,\mathcal{NN},\theta}_{\tau_j,\tau_{j+1}} \bar{v}^{(m),n}_{\tau_{j+1}}(x) = (A^{\theta, \sum_{i=1}^{s-1} d_i(d_{i-1}+1)}_{d_{s-1},1} \circ \mathcal{L}^{\text{ReLU}}_{d_{s-1}} \circ A^{\theta, \sum_{i=1}^{s-2} d_i(d_{i-1}+1)}_{d_{s-2},d_{s-1}}$$
$$\circ \cdots \circ \mathcal{L}^{\text{ReLU}}_{d_2} \circ A^{\theta, d_1(d_0+1)}_{d_1,d_2} \circ \mathcal{L}^{\text{ReLU}}_{d_1} \circ A^{\theta,0}_{d_0,d_1})(x), \quad x \in \mathbb{R}^N$$

with $\theta^*_{\tau_j}$ given by $\theta^*_{\tau_j} = \operatorname{argmin}_\theta L^{\tau_j}(\theta)$ with the loss function $L^{\tau_j} : \mathbb{R}^\nu \to \mathbb{R}$:

$$L^{\tau_j}(\theta) = \mathbb{E}\bigg[\bigg| Q^{(m),n,\mathcal{NN},\theta}_{\tau_j,\tau_{j+1}} \bar{v}^{(m),n,\mathcal{NN}}_{\tau_{j+1}}(\bar{X}^{\text{EM},\tau_0,x}_{\tau_j})$$
$$- (Q^{(m),n,\mathcal{NN},\theta^*_{\tau_{j+1}}}_{\tau_{j+1},\tau_{j+2}} \bar{v}^{(m),n,\mathcal{NN}}_{\tau_{j+2}} \vee f(\tau_{j+1}, \cdot))(\bar{X}^{\text{EM},\tau_0,x}_{\tau_{j+1}}) \prod_{\ell=1}^n W^{(m),t_{jn+\ell-1}, \bar{X}^{\text{EM},\tau_0,x}_{t_{jn+\ell-1}}}_{t_{jn+\ell}} \bigg|^2 \bigg].$$

The algorithm of the scheme is given as follows.

Algorithm 6.1 Deep learning-based backward dynamic programming algorithm with weak approximation

Require: $M \in \mathbb{N}$ (batch size), $J \in \mathbb{N}$ (number of train-steps), $\gamma \in (0, 1)$ (learning rate), $K \in \mathbb{N}$ (number of exercise dates), $n \in \mathbb{N}$ (number of time steps for each interval of the early-exercise dates)

1: **for** $i = K - 1$ to 0 **do**
2:   **for** $j = 1$ to J **do**
3:     **for** $r = 1$ to M **do**
4:       $\bar{X}^{\text{EM},\tau_0,x,r,j}_{\tau_0} = x$,
5:       **for** $\ell = 1$ to $(i+1)n$ **do**
6:         Generate i.i.d. Gaussian RVs $\Delta W^{1,r,j}_{t_{\ell-1},t_\ell}, \ldots, \Delta W^{d,r,j}_{t_{\ell-1},t_\ell}$
7:         $\bar{X}^{\text{EM},\tau_0,x,r,j}_{t_\ell}$
$= \bar{X}^{\text{EM},\tau_0,x,r,j}_{t_{\ell-1}} + \sigma_0(\bar{X}^{\text{EM},\tau_0,x,r,j}_{t_{\ell-1}}) + \sum_{e=1}^d \sigma_e(\bar{X}^{\text{EM},\tau_0,x,r,j}_{t_{\ell-1}}) \Delta W^{e,r,j}_{t_{\ell-1},t_\ell}$
8:       **end for**
9:       Compute $\mathcal{W}^{(m),t_{in+\ell-1},\bar{X}^{\text{EM},\tau_0,x,r,j}_{t_{in+\ell-1}}}_{t_{in+\ell}}$ for $\ell = 1, \ldots, n$
10:     **end for**
11:     **if** $i = K - 1$ **then**
12:       $Q^{(m),n,\mathcal{NN},\Theta_{\tau_{i+1}}}_{\tau_{i+1},\tau_{i+2}} \bar{v}^{(m),n,\mathcal{NN}}_{\tau_{i+2}}(\cdot) = f(\tau_{i+1}, \cdot)$
13:     **end if**
14:     **if** $i = 0$ **then**
15:       $L^{\tau_0,j}(\theta^j_0) = \frac{1}{M} \sum_{r=1}^M \big[\theta^j_0$
$- (Q^{(m),n,\mathcal{NN},\Theta_{\tau_1}}_{\tau_1,\tau_2} \bar{v}^{(m),n,\mathcal{NN}}_{\tau_2} \vee f(\tau_1, \cdot))(\bar{X}^{\text{EM},\tau_0,x,r,j}_{\tau_1}) \prod_{\ell=1}^n \mathcal{W}^{(m),t_\ell-1,\bar{X}^{\text{EM},\tau_0,x,r,j}_{t_{\ell-1}}}_{t_\ell} \big]^2$,
16:     **else**
17:       $L^{\tau_i,j}(\theta^j_{\tau_i}) = \frac{1}{M} \sum_{r=1}^M \big[Q^{(m),n,\mathcal{NN},\theta^j_{\tau_i}}_{\tau_i,\tau_{i+1}} \bar{v}^{(m),n,\mathcal{NN}}_{\tau_{i+1}}(\bar{X}^{\text{EM},\tau_0,x,r,j}_{\tau_i})$
$- (Q^{(m),n,\mathcal{NN},\Theta_{\tau_{i+1}}}_{\tau_{i+1},\tau_{i+2}} \bar{v}^{(m),n,\mathcal{NN}}_{\tau_{i+2}} \vee f(\tau_{i+1}, \cdot))(\bar{X}^{\text{EM},\tau_0,x,r,j}_{\tau_{i+1}}) \prod_{\ell=1}^n \mathcal{W}^{(m),t_{in+\ell-1},\bar{X}^{\text{EM},\tau_0,x,r,j}_{t_{in+\ell-1}}}_{t_{in+\ell}} \big]^2$,
18:     **end if**
19:     where $Q^{(m),n,\mathcal{NN},\theta}_{\tau_i,\tau_{i+1}} \bar{v}^{(m),n,\mathcal{NN}}_{\tau_{i+1}}$ is a function given by neural network
20:     Minimize the loss function $L^{\tau_i,j}(\theta^j_{\tau_i})$ and obtain the parameter $\theta^{*j}_{\tau_i}$ using ADAM optimizer
21:     Update the parameter $\theta^{j+1}_{\tau_i} = \theta^{*j}_{\tau_i}$
22:   **end for**
23:   $\Theta_{\tau_i} = \theta^J_{\tau_i}$
24: **end for**
25: Return $\bar{v}^{(m),n,n,\mathcal{NN}}_0(x) = \Theta_0 \vee f(\tau_0, x)$

Remark 6.1 In the above algorithm, the optimized parameter $\theta^*_{\tau_i}$ is obtained by using stochastic gradient descent (SGD) J-times repeatedly with M-paths of the SDE, for each $i = K - 1, \ldots, 1$. In the numerical experiments below, we employ Adam (see Kingma and Ba 2015) and batch normalization (see Ioffe and Szegedy 2015) as SGD to accelerate the learning processes.

Below, we show numerical examples where experiments are performed in Python with Tensorflow 1.15 on two GPUs (NVIDIA GA102GL [RTX A6000] 48GB with NVLink) where the underlying system consists of an AMD EPYC 7402P 2.80GHz 24 cores CPU with 128 GB DDR4-3200 memory running on Ubuntu 18.04.

6.6 Algorithm, Implementation and Numerical Results

Table 6.1 Numerical result for 100-dimensional SABR model

Scheme	n	Error	Standard deviation	Computing Time
$m=1$	2^7	0.0012	0.0124	1171.5 s
$m=2$	2^2	0.0007	0.0113	150.8 s

Let $W = \{W_t\}_{0 \leq t \leq T}$ be $2d$-dimensional Brownian motion. We consider the following $2d$-dimensional log-normal SABR model (d-asset price and their d-volatility processes):

$$dS_t^i = rS_t^i dt + \sigma_t^i S_t^i dW_t^{2i-1}, \quad S_0^i = x_{2i-1} > 0, \quad i = 1, \ldots, d$$
$$d\sigma_t^i = \nu\sigma_t^i(\rho dW_t^{2i-1} + \sqrt{1-\rho^2}dW_t^{2i-1}), \quad \sigma_0^i = x_{2i} > 0, \quad i = 1, \ldots, d,$$

with $\nu > 0$, $\rho \in (-1, 1)$. We consider the Bermudan put option with the payoff function $f(t, x) = e^{-rt}(0.0 \vee (70.0 - x_1) \vee (70.0 - x_3) \vee \cdots \vee (70.0 - x_{2d-1}))$ for $x \in \mathbb{R}^{2d}$.

First, we perform numerical experiment with the case of $2d = 2$. We set the parameters as $K = 4$ (the number of early exercise dates), $T = 1.0, r = 0.02, x_1 = 100.0$, $x_2 = 0.25, \nu = 0.1, \rho = -0.4$. We implement the algorithm by using neural networks with 1 input layer with 2-neurons, two hidden layer with 12-neurons and 1 output layer with 1-neuron with ReLU activation function. The hyperparameters of the neural networks are set as $M = 16384$ (batch size), $J = 4000$ (training steps) and $\gamma(j) = 10^{-2}\mathbf{1}_{[0.0,0.3J]}(j) + 10^{-3}\mathbf{1}_{(0.3J,0.6J]}(j) + 10^{-4}\mathbf{1}_{(0.6J,J]}(j)$ (learning rate). We estimate the approximation error of each scheme by $|v^{\text{ref}}(0, x) - v^{(m)}(0, x)|$, $m = 1, 2, 3$ where the approximation values of the Bermudan option price $v^{(m)}(0, x)$, $m = 1, 2, 3$ are computed by the average of 25 independent runs of the algorithm of order m and the reference value $v(0, x)$ is computed by 100 independent runs of the neural network algorithm based on the algorithm of order $m = 1$ under the same neural network settings with $n = 256$-times discretization for each interval between the early exercise dates (i.e., we discretize one year $T = 1$ with $(n \times K) = 1024$-equidistant grids in total). The approximation error for each scheme is plotted in Fig. 6.1.

Next, we consider higher dimensional case ($2d = 100$). The parameters are set as $K = 4$, $T = 1.0$, $r = 0.02$, $(x_1, x_3, \ldots, x_{99}) = (100.0, \ldots, 100.0)$, $(x_2, x_4, \ldots, x_{100}) = (0.15, \ldots, 0.15)$, $\nu = 0.1, \rho = -0.5$. We implement the algorithm by using neural networks with 1 input layer with 100-neurons, two hidden layer with 110-neurons and 1 output layer with 1-neuron with ReLU activation function. The hyperparameters of the neural networks are set as $M = 16384$ (batch size), $J = 4000$ (training steps) and $\gamma(j) = 10^{-1}\mathbf{1}_{[0.0,0.3J]}(j) + 10^{-2}\mathbf{1}_{(0.3J,0.6J]}(j) + 10^{-3}\mathbf{1}_{(0.6J,J]}(j)$ (learning rate). We estimate the approximation error of the algorithm of order $m = 1, 2$ in the same manner as in the previous example. The result is summarized in Fig. 6.2 and Table 6.1.

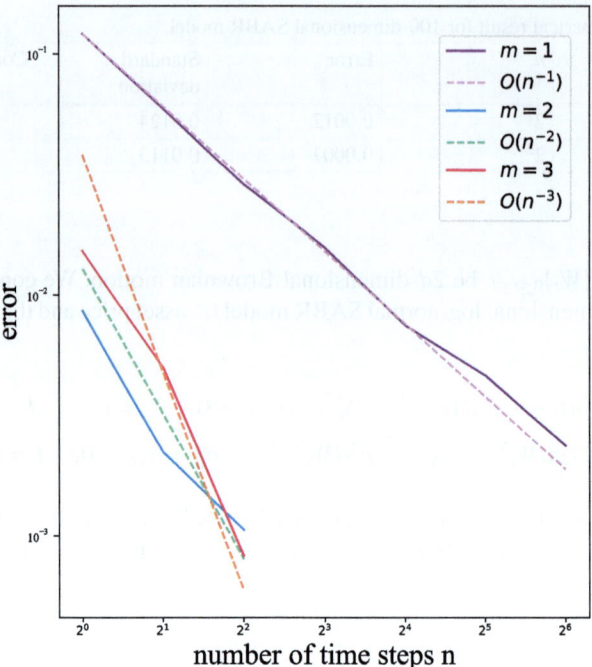

Fig. 6.1 SABR 2-dim

6.7 Notes and Summary

Firstly, we explained that the solution of a linear PDE $(\partial_t + \mathcal{L})u(t, x)$, $u(T, \cdot) = f$ is approximated by a weak approximation function $(Q^{(m)}_{(T-t)/n})^n f(\cdot)$ and a neural network $\phi_{\mathcal{N}\mathcal{N}}$ as $u(t, \cdot) = P_{T-t} f \approx (Q^{(m)}_{(T-t)/n})^n f \approx \phi_{\mathcal{N}\mathcal{N}}$ through a loss function obtained by the representation:

$$[[a, b]^N \ni x \mapsto (Q^{(m)}_{(T-t)/n})^n f(x)]$$
$$= \mathrm{argmin}_{v \in C([a,b]^N)} \mathbb{E}[|v(\xi)$$
$$- f(\bar{X}^{\mathrm{EM},\xi,(n)}_{T-t}) \prod_{i=1}^{n} \mathcal{M}^{(m)}((T-t)/n, \bar{X}^{\mathrm{EM},\xi,(n)}_{(i-1)(T-t)/n}, W_{i(T-t)/n} - W_{(i-1)(T-t)/n})|^2]$$

with an independent random variable $\xi \in U([a, b]^N)$. Then, we showed that the method can be applied to a nonlinear problem. That is, for an optimal stopping problem:

$$v_{\tau_i}(\cdot) = \sup_{\tau \in \mathcal{T}_i} \mathbb{E}[f(\tau, X_\tau)|X_{\tau_i} = \cdot], \quad i = 0, 1, \ldots, K,$$

6.7 Notes and Summary

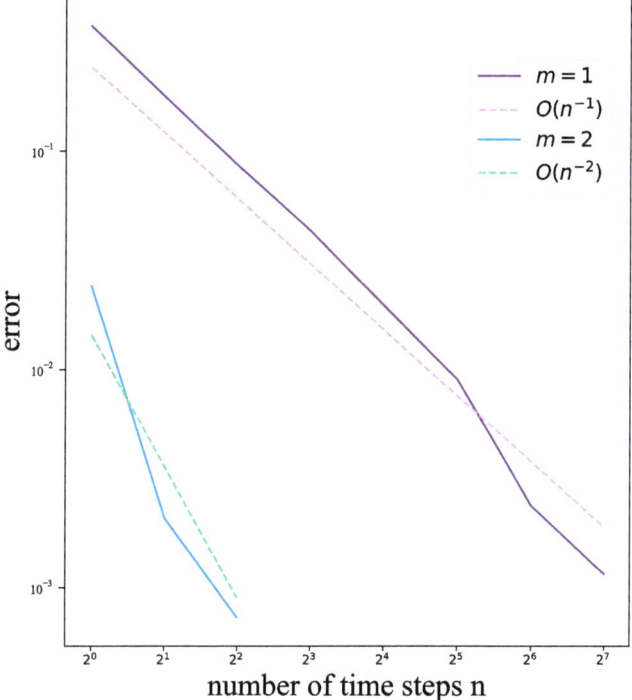

Fig. 6.2 SABR 100-dim

a deep learning-based backward dynamic programming algorithm is constructed based on Naito and Yamada (2024b) using a weak approximation scheme as follows:

$$\bar{v}_{\tau_K}^{(m),n,\mathcal{NN}}(\cdot) = f(\tau_K, \cdot),$$
$$\bar{v}_{\tau_j}^{(m),n,\mathcal{NN}}(\cdot) = (Q_{\tau_j,\tau_{j+1}}^{(m),n,\mathcal{NN}} \bar{v}_{\tau_{j+1}}^{(m),n,\mathcal{NN}})(\cdot) \vee f(\tau_j, \cdot), \ j = K-1,\ldots,1,$$
$$\bar{v}_{\tau_0}^{(m),n,\mathcal{NN}}(\cdot) = (Q_{\tau_0,\tau_1}^{(m),n} \bar{v}_{\tau_1}^{(m),n,\mathcal{NN}})(\cdot) \vee f(\tau_0, \cdot),$$

with neural networks $Q_{\tau_j,\tau_{j+1}}^{(m),n,\mathcal{NN}} \bar{v}_{\tau_{j+1}}^{(m),n,\mathcal{NN}}$, $j = K-1,\ldots,1$ such that

$$\|Q_{\tau_j,\tau_{j+1}}^{(m),n,\mathcal{NN}} \bar{v}_{\tau_{j+1}}^{(m),n,\mathcal{NN}} - Q_{\tau_j,\tau_{j+1}}^{(m),n} \bar{v}_{\tau_{j+1}}^{(m),n,\mathcal{NN}}\|_{L^2(\mathbb{R}^N, \mathscr{B}(\mathbb{R}^N), P \circ (\bar{X}_{\tau_j}^{EM,\tau_0,x})^{-1})} \leq K^{-1} n^{-m},$$

where

$$Q_{\tau_j,\tau_{j+1}}^{(m),n} g = \mathbb{E}\left[g(\bar{X}_{\tau_{j+1}}^{EM,\tau_i,x}) \prod_{\ell=1}^{n} \mathcal{W}_{t_{jn+\ell}}^{(m),t_{jn+\ell-1},\bar{X}_{t_{jn+\ell-1}}^{EM,\tau_i,x}} | \bar{X}_{\tau_j}^{EM,\tau_i,x} = \ \cdot\ \right]$$

with $\mathcal{W}_{t_{i+1}}^{(m),t_i,x} = \mathcal{M}^{(m)}(t_{i+1}-t_i, x, W_{t_{i+1}} - W_{t_i})$ and $t_{jn+\ell} = \tau_j + (\tau_{j+1}-\tau_j)\ell/n$, which provides $v_{\tau_0}(\cdot) = \bar{v}_{\tau_0}^{(m),n,\mathcal{NN}}(\cdot) + O(n^{-m})$. Here, the existence of such neural network $Q_{\tau_j,\tau_{j+1}}^{(m),n,\mathcal{NN}} \bar{v}_{\tau_{j+1}}^{(m),n,\mathcal{NN}}$ is justified by a version of the universal approximation theorem (Theorem 6.1) introduced in this chapter. Numerical results showed that the scheme gives an efficient approximation even when the dimension of models is high. Since such objective appears in other nonlinear problems, the scheme can be applied to backward stochastic differential equations (BSDEs). In summary, the deep learning-based weak approximation scheme will be an alternative of the least squares Monte-Carlo method for nonlinear problems. See the related studies on deep learning approaches such as Beck et al. (2021a), Gonon et al. (2021), Iguchi et al. (2021), Naito and Yamada (2022a), Takahashi and Yamada (2023) for weak approximation of SDEs or linear PDEs, and Huré et al. (2020), Beck et al. (2021b), Naito and Yamada (2022b), Naito and Yamada (2024a) for splitting for BSDEs or nonlinear PDEs, for instance.

We comment on other deep learning methods for solving BSDEs or nonlinear PDEs combining with asymptotic expansion and weak approximation. After Han and Jentzen (2017) and Han et al. (2018) introduced the deep BSDE method which solves BSDEs by an efficient stochastic optimization problem with a deep learning technique, Fujii et al. (2019) proposed a deep learning-based scheme for solving BSDEs using an asymptotic expansion as a prior knowledge for constructing a neural network. Also, Naito and Yamada (2020) provided a deep BSDE solver with a weak approximation scheme and Takahashi et al. (2022) provided a control variate method using a decomposition of BSDE inspired by the asymptotic expansion of Takahashi and Yamada (2015b). See Chessari et al. (2023) and references therein.

Constructing a deep learning or AI method for solving high-dimensional BSDEs or nonlinear PDEs is an important topic and still has open problems in applied mathematics and other fields. We believe that efficient methods will be proposed and applied to various problems in industries.

References

R. Arora, A. Basu, P. Mianjy, A. Mukherjee, Understanding deep neural networks with rectified linear units, in *ICLR* (2018)

V. Bally, D. Talay, The law of the Euler scheme for stochastic differential equations I: Convergence rate of the distribution function. Probab. Theory Relat. Fields **104**, 43–60 (1996)

V. Bally, L. Caramellino, R. Cont, *Stochastic Integration by Parts and Functional Itô Calculus* (Birkhäuser, 2016)

C. Bayer, P. Friz, R. Loeffen, Semi-closed form cubature and applications to financial diffusion models. Quantitat. Financ. **13**(5), 769–782 (2013)

C. Beck, S. Becker, P. Grohs, N. Jaafari, A. Jentzen, Solving stochastic differential equations and Kolmogorov equations by means of deep learning. J. Sci. Comput. 3 (2021a)

C. Beck, S. Becker, P. Cheridito, A. Jentzen, A. Neufeld, Deep splitting method for parabolic PDEs. SIAM J. Sci. Comput. **43**(5), 3135–3154 (2021b)

O. Calin, *Deep Learning Archtechture* (Springer, 2020)

J. F. Carriere, Valuation of early-exercise price of options using simulations and nonparametric regression. Insuran.: Math. Econ. **19**, 19–30 (1996)

J. Chessari, R. Kawai, Y. Shinozaki, T. Yamada, Numerical methods for backward stochastic differential equations: A survey. Probab. Surv. **20**, 486–567 (2023)

W.E, J. Han, A. Jentzen, Deep learning-based numerical methods for high dimensional parabolic partial differential equations and backward stochastic differential equations. Communications in Mathematics and Statistics **5**(4) 349–380 (2017)

A. Friedman, *Partial Differential Equations of Parabolic Type* (Prentice Hall, 1964)

P. Friz, M. Hairer, *A Course on Rough Paths* (Springer, 2014)

P. Friz, N. Victoir, *Multidimensional Stochastic Processes as Rough Paths: Theory and Applications* (Cambridge University Press, 2009)

M. Fujii, A. Takahashi, Solving backward stochastic differential equations with quadratic-growth drivers by connecting the short-term expansions. Stochast. Process. Their Appl. **129**(5), 1492–1532 (2019)

M. Fujii, A. Takahashi, M. Takahashi, Asymptotic expansion as prior knowledge in deep learning method for high dimensional BSDEs. Asia-Pacific Finan. Markets. **26**(3), 391–408 (2019)

T. Funaki, *Stochastic Differential Equation* (in Japanese) (Iwanami, 1997)

E. Gobet, *Monte-Carlo Methods and Stochastic Processes: From Linear to Non-Linear* (Chapman and Hall/CRC, 2020)

L. Gonon, P. Grohs, A. Jentzen, D. Kofler, D. Siska, Uniform error estimates for artificial neural network approximations for heat equations. IMA J. Numer. Anal. **42**(3), 1991–2054 (2021)

J. Han, A. Jentzen, E. Weinan, Solving high-dimensional partial differential equations using deep learning. Proc. Natl. Acad. Sci. **115**(34), 8505–8510 (2018)

K. Hornik, M. Stinchcombe, H. White, Multilayer feedforward networks are universal approximators. Neural Netw. **2**, 359–366 (1989)

K. Hornik, M. Stinchcombe, H. White, Universal approximation of an unknown mapping and its derivatives using Multilayer feedforward networks. Neural Netw. **3**, 551–560 (1990)

K. Hornik, Approximation capabilities of Multilayer feedforward networks. Neural Netw. **4**, 251–257 (1991)

C. Huré, H. Pham, X. Warin, Deep backward schemes for high-dimensional nonlinear PDEs. Math. Comput. **89**(324), 1547–1579 (2020)

N. Ikeda, S. Watanabe, *Stochastic Differential Equations and Diffusion Processes*, 2nd edn. (North-Holland, Amsterdam, 1989)

Y. Iguchi, A. Beskos, Parameter inference for hypo-elliptic diffusions under a weak design condition. Electron. J. Statist. **19**(1), 1337–1369 (2025)

Y. Iguchi, A. Beskos, M. Graham, Parameter inference for degenerate diffusion processes. Stochast. Process. Their Appl. **174**, 104384 (2024a)

Y. Iguchi, A. Beskos, M. Graham, Parameter estimation with increased precision for elliptic and hypo-elliptic diffusions. Bernoulli **31**(1), 333–358 (2024b)

Y. Iguchi, A. Jasra, M. Maama, A. Beskos, Antithetic multilevel methods for elliptic and hypo-elliptic diffusions with applications. SIAM/ASA Journal on Uncertainty Quantification **13**(2), 805–830 (2025)

Y. Iguchi, R. Naito, Y. Okano, A. Takahashi, T. Yamada, Deep asymptotic expansion: application to financial mathematics, in *IEEE CSDE 2021* (2021)

Y. Iguchi, T. Yamada, A second order discretization for degenerate systems of stochastic differential equations. IMA J. Numer. Anal. **41**(4), 2782–2829 (2020)

Y. Iguchi, T. Yamada, Operator splitting around Euler-Maruyama scheme and high order discretization of heat kernels. ESAIM: Math. Modell. Numer. Anal. **55**, 323–367 (2021)

Y. Iguchi, T. Yamada, Weak approximation of SDEs for tempered distributions and applications. Adv. Comput. Math. **48**(5), 52 (2022)

S. Ioffe, C. Szegedy, Batch normalization: accelerating deep network training by reducing internal covariate shift, in *Proceedings of the 32nd International Conference on Machine Learning*, vol. 37 (2015), pp. 448–456

K. Itô, Differential equations determining Markov processes (in Japanese). Zenkoku Shijo Sugaku Danwakai **1077**, 1352–1400 (1942)

J. Jacod, P. Protter, *Probability Essentials* (Springer, 2004)

I. Karatzas, S. Shreve, *Brownian Motion and Stochastic Calculus*, 2nd edn. (Springer, 1991)

D. Kingma, J. Ba, Adam: a method for stochastic optimization, in *Proceedings of the International Conference on Learning Representations (ICLR)* (2015)

P.E. Kloeden, E. Platen, *Numerical Solution of Stochastic Differential Equations* (Springer, 1992)

S. Kusuoka, Approximation of expectation of diffusion process and mathematical finance. Adv. Stud. Pure Math. **31**, 147–165 (2001)

S. Kusuoka, Malliavin calculus revisited. J. Math. Sci. Univ. Tokyo **10**(2), 261–277 (2003)

S. Kusuoka, Approximation of expectation of diffusion processes based on Lie algebra and Malliavin calculus, in *Advances in Mathematical Economics* (2004), pp. 69–83

S. Kusuoka, *Stochastic Analysis* (Springer, 2020)

S. Kusuoka, D. Stroock, Applications of the Malliavin calculus Part I, in *Stochastic Analysis (Katata/Kyoto 1982)* (1984), pp. 271–306

S. Kusuoka, D. Stroock, Applications of the Malliavin calculus Part II. J. Fac. Sci. Univ. Tokyo **1**, 1–76 (1985)

H. Kunita, *Stochastic Flows and Stochastic Differential Equations* (Cambridge University Press, 1990)

N. Kunitomo, A. Takahashi, Pricing average options. Japan Financ. Rev. **14**, 1–20 (in Japanese) (1992)

N. Kunitomo, A. Takahashi, The asymptotic expansion approach to the valuation of interest rate contingent claims. Math. Financ. **11**, 117–151 (2001)

N. Kunitomo, A. Takahashi, On validity of the asymptotic expansion approach in contingent claim analysis. Ann. Appl. Prob. **13**(3), 914–952 (2003)

F.A. Longstaff, E.S. Schwartz, Valuing American options by simulation: a simple least-squares approach. Rev. Financ. Studies **14**(1), 113–147 (2001)

T. Lyons, N. Victoir, Cubature on Wiener space, Proceedings of the Royal Society of London A: Mathematical. Phys. Eng. Sci. **460**, 169–198 (2004)

P. Malliavin, *Integration and Probability* (Springer, 1995)

P. Malliavin, *Stochastic Analysis* (Springer, 1997)

P. Malliavin, A. Thalmaier, *Stochastic Calculus of Variations in Mathematical Finance* (Springer, 2006)

Matsumoto and Taniguchi, *Stochastic Analysis: Itô and Malliavin Calculus in Tandem* (Cambridge University Press, 2016)

Y. Morimoto, M. Sasada, Algebraic structure of vector fields in financial diffusion models and its applications. Quantit. Financ. **17**(7), 1105–1117 (2017)

R. Naito, T. Yamada, A third-order weak approximation of multidimensional Itô stochastic differential equations. Monte Carlo Methods Appl. **25**(2), 97–120 (2019)

R. Naito, T. Yamada, A second-order discretization for forward-backward SDEs using local approximations with Malliavin calculus. Monte Carlo Methods Appl. **25**(4), 341–361 (2019)

R. Naito, T. Yamada, An acceleration scheme for deep learning-based BSDE solver using weak expansions. Int. J. Financ. Engin. **7**(2), 2050012 (2020)

R. Naito, T. Yamada, A higher order weak approximation of Mckean-Vlasov type SDEs. BIT Numer. Math. **62**, 521–559 (2021)

R. Naito, T. Yamada, Deep weak approximation of SDEs: a spatial approximation scheme for solving Kolmogorov equations. Int. J. Comput. Methods **19**(8), 2142014 (2022)

R. Naito, T. Yamada, A deep learning-based high-order operator splitting method for high dimensional nonlinear parabolic PDEs via Malliavin calculus: application to CVA computation, *IEEE CIFEr* (2022b), pp. 1–8

R. Naito, T. Yamada, Deep high order splitting method for semilinear degenerate PDEs and application to high dimensional nonlinear pricing models. Digit. Financ. **6**, 693–725 (2024)

R. Naito, T. Yamada, Pricing high-dimensional Bermudan options using deep learning and high-order weak approximation. J. Comput. Financ. **28**(1), 65–94 (2024b)

S. Ninomiya, N. Victoir, Weak approximation of stochastic differential equations and application to derivative pricing. Appl. Math. Financ. **15**(2), 107–121 (2008)

D. Nualart, *The Malliavin Calculus and Related Topics* (Springer, 2006)

D. Nualart, E. Nualart, *Introduction to Malliavin Calculus* (Cambridge University Press, 2018)

G. Pages, *Numerical Probability* (Springer, 2018)

L.C.G. Rogers, D. Williams, *Diffusions, Markov Processes and Martingales Volume I, II*, 2nd edn. (Wiley, 1994)

W. Rudin, *Real and Complex Analysis* (McGraw-Hill Publishing Co., 1987)

I. Shigekawa, Stochastic Analysis (American Mathematical Society, 2004)

Y. Shinozaki, Construction of a third-order K-scheme and its application to financial models. SIAM J. Financ. Math. **8**(1), 901–932 (2017)

K. Shiraya, A. Takahashi, T. Yamada, Pricing discrete barrier options under stochastic volatility. Asia-Pacific Finan. Markets. **19**(3), 205–232 (2012)

D. Stroock, Homogeneous chaos revisited. Seminaire de Probabilités XXI **1247**, 1–8 (1987)

D. Stroock, *Partial Differential Equations for Probabilists* (Cambridge University Press, 2008)

A. Takahashi, Essays on the Valuation Problems of Contingent Claims, Ph.D.Dissertation, Haas School of Business, University of California, Berkeley (1995)

A. Takahashi, An asymptotic expansion approach to pricing financial contingent claims. Asia-Pacific Finan. Markets. **6**(2), 115–151 (1999)

A. Takahashi, Asymptotic expansion approach in finance, *Large Deviations and Asymptotic Methods in Finance*, Springer Proceedings in Mathematics & Statistics, ed. by P. Friz, J. Gatheral, A. Gulisashvili, A. Jacquier, J. Teichmann (2015), pp. 345–411

A. Takahashi, Y. Tsuchida, T. Yamada, A new efficient approximation scheme for solving high-dimensional semilinear PDEs: control variate method for Deep BSDE solver. J. Comput. Phys. **454**, 110956 (2022)

A. Takahashi, T. Yamada, An asymptotic expansion with push-down of Malliavin weights. SIAM J. Financ. Math. **3**, 95–136 (2012)

A. Takahashi, T. Yamada, On error estimates for asymptotic expansions with Malliavin weights: application to stochastic volatility model. Math. Oper. Res. **40**(3), 513–551 (2015)

A. Takahashi, T. Yamada, An asymptotic expansion of forward-backward SDEs with a perturbed driver. Int. J. Financ. Engin. **2**(2), 1550020 (2015b)

A. Takahashi, T. Yamada, A weak approximation with asymptotic expansion and multidimensional Malliavin weights. Ann. Appl. Prob. **26**(2), 818–856 (2016)

A. Takahashi, T. Yamada, Solving Kolmogorov PDEs without the curse of dimensionality via deep learning and asymptotic expansion with Malliavin calculus. Partial Differ. Equ. Appl. **4**, article number 27 (2023)

A. Takahashi, T. Yamada, New asymptotic expansion formula via Malliavin calculus and its application to rough differential equation driven by fractional Brownian motion. Asymptot. Anal. **140**(1–2), 37–58 (2024)

A. Takahashi, N. Yoshida, An asymptotic expansion scheme for optimal investment problems. Stat. Infer. Stoch. Process. **7**(2), 153–188 (2004)

A. Takahashi, N. Yoshida, Monte Carlo simulation with asymptotic method. J. Jpn. Stat. Soc. **35**(2), 171–203 (2005)

J.N. Tsitsiklis, B. Van Roy, Regression methods for pricing complex American-style options. IEEE Trans. Neural Netw. **12**(4), 694–703 (2001)

A. S. Üstünel, *An Introduction to Analysis on Wiener Space* (Springer, 1995)

S. Wang, X. Sun, Generalization of hinging hyperplanes. IEEE Trans. Inf. Theory **51**(12), 4425–4431 (2005)

S. Watanabe, *Lectures on Stochastic Differential Equations and Malliavin Calculus* (Springer, 1984)

S. Watanabe, Analysis of Wiener functionals (Malliavin calculus) and its applications to heat kernels. Ann. Probab. **15**, 1–39 (1987)

N. Wiener, Differential space. J. Math. Phys. **58**, 131–174 (1923)

D. Williams, *Probability with Martingale* (Cambridge University Press, 1991)

T. Yamada, Weak Milstein scheme without commutativity condition and its error bound. Appl. Numer. Math. **131**, 95–108 (2018)

T. Yamada, An arbitrary high order weak approximation of SDE and Malliavin Monte Carlo: application to probability distribution functions. SIAM J. Numer. Anal. **57**(2), 563–591 (2019)

T. Yamada, High order weak approximation for irregular functionals of time-inhomogeneous SDEs. Monte Carlo Methods Appl. **25**(2), 97–120 (2021)

T. Yamada, A Gaussian Kusuoka-approximation without solving random ODEs. SIAM J. Financ. Math. **13**(1), SC1-SC11 (2022)

T. Yamada, A new algorithm for computing path integrals and weak approximation of SDEs inspired by large deviations and Malliavin calculus. Appl. Numer. Math. **187**, 192–205 (2023)

T. Yamada, Total variation bound for Milstein scheme without iterated integrals. Monte Carlo Methods Appl. **29**(3), 221–242 (2023)

T. Yamada, K. Yamamoto, A second order weak approximation of SDEs using Markov chain without Lévy area simulation. Monte Carlo Methods Appl. **24**(4), 289–308 (2018)

T. Yamada, K. Yamamoto, Second order discretization of Bismut-Elworthy-Li formula: application to sensitivity analysis. SIAM/ASA J. Uncertainty Quantif. **7**(1), 143–173 (2019)

T. Yamada, K. Yamamoto, A second-order discretization with Malliavin weight and Quasi-Monte Carlo method for option pricing. Quant. Financ. **20**(11), 1825–1837 (2020)

N. Yoshida, Asymptotic expansions for statistics related to small diffusions. J. Jpn. Stat. Soc. **22**, 139–159 (1992)

N. Yoshida, Asymptotic expansion for small diffusions via the theory of Malliavin-Watanabe. Probab. Theory Relat. Fields **92**, 275–311 (1992)

MIX
Papier aus verantwortungsvollen Quellen
Paper from responsible sources
FSC® C105338

If you have any concerns about our products,
you can contact us on
ProductSafety@springernature.com

In case Publisher is established outside the EU,
the EU authorized representative is:
Springer Nature Customer Service Center GmbH
Europaplatz 3, 69115 Heidelberg, Germany

Printed by Libri Plureos GmbH
in Hamburg, Germany